T0052039

FIND
MORE
BIRDS

A Chestnut-sided Warbler during spring migration in Brooklyn, NY

FIND MORE BIRDS

111 SURPRISING WAYS TO SPOT BIRDS WHEREVER YOU ARE

Written and photographed by

HEATHER WOLF

THE EXPERIMENT
NEW YORK

For Connor: Kickee-doo!

FIND MORE BIRDS: *111 Surprising Ways to Spot Birds Wherever You Are*
Text and photographs, including cover photographs, copyright © 2023 by Heather Wolf
No audio playback or flash was used to obtain the photographs in this book.

All rights reserved. Except for brief passages quoted in newspaper, magazine, radio, television, or online reviews, no portion of this book may be reproduced, distributed, or transmitted in any form or by any means, electronic or mechanical, including photocopying, recording, or information storage or retrieval system, without the prior written permission of the publisher.

The Experiment, LLC
220 East 23rd Street, Suite 600
New York, NY 10010-4658
theexperimentpublishing.com

THE EXPERIMENT and its colophon are registered trademarks of The Experiment, LLC. Many of the designations used by manufacturers and sellers to distinguish their products are claimed as trademarks. Where those designations appear in this book and The Experiment was aware of a trademark claim, the designations have been capitalized.

The Experiment's books are available at special discounts when purchased in bulk for premiums and sales promotions as well as for fund-raising or educational use. For details, contact us at info@theexperimentpublishing.com.

Library of Congress Cataloging-in-Publication Data

Names: Wolf, Heather, author.
Title: Find more birds : 111 surprising ways to spot birds wherever you are / Heather Wolf.
Description: New York : The Experiment, [2023] | Includes bibliographical
   references.
Identifiers: LCCN 2023005639 (print) | LCCN 2023005640 (ebook) | ISBN
   9781615199402 (paperback) | ISBN 9781615199419 (ebook)
Subjects: LCSH: Bird watching. | Bird watching--Guidebooks.
Classification: LCC QL677.5 .W64 2023 (print) | LCC QL677.5 (ebook) | DDC
   598.072/34--dc23/eng/20230206
LC record available at https://lccn.loc.gov/2023005639
LC ebook record available at https://lccn.loc.gov/2023005640

ISBN 978-1-61519-940-2
Ebook ISBN 978-1-61519-941-9

Cover and text design by Jack Dunnington
Author photograph by Harold Moeller

Manufactured in China

First printing August 2023
10 9 8 7 6 5 4 3 2 1

CONTENTS

INTRODUCTION 4

1 BIRD-FINDING BASICS

1. Follow sounds 12
2. Look for movement 14
3. Check freshwater sources, even drips 16
4. Stay in one spot for at least 10 minutes 18
5. Look during migration 20
6. Get a bird field guide . . . made of paper 22
7. Get bird-worthy binoculars 24
8. Scan the sky 26
9. Take a seat 28
10. Avoid sudden movements and remain calm 30
11. Look for birdlike shapes and silhouettes 32
12. Find the high point 36
13. Don't lose hope if it flies away 38
14. Change your route 40
15. Stay fueled—always have snacks on hand 42
16. Dress for success 44
17. Avoid stressing birds out— don't play recordings 46
18. Pick a patch 48
19. Keep a list 50

2 FINDING BIRDS AT THEIR FAVORITE RESTAURANTS

20. Discover hidden gems in leaf litter 54
21. Scan the water . . . twice or thrice 56
22. Search for seeds 60
23. Check around the edges 62
24. Hang out at the visitor center or parking lot 64
25. Visit gardens with native plants 66
26. Beware of bark that moves 70
27. Watch mudflats and other muddy patches 72
28. Look for birds on ball fields, even with artificial turf 74
29. Check for open patches of ground after a snowfall 76
30. Watch fish ponds large and small 78
31. Visit berry buffets 80
32. Creep up on cattails 82
33. Visit bird feeders 84
34. Visit a busy urban park 86
35. Bundle up and hang out near the jetties 88
36. Visit a beautified dump 90
37. Walk around your local college campus 92

FINDING BIRDS BY THE CLUES THEY LEAVE

38. Check for cavities 96
39. Check above and around bird poop 98
40. Smell the scent of fallen flowers, and check them for slits! 100
41. Look out for bird spas . . . dust baths 102
42. Follow tracks 104
43. Know your woodpecker drill holes 106
44. Look for fallen feathers, a sign of nearby predators 108

 FINDING BIRDS THROUGH TECH SUPPORT

45. Let Merlin listen 112
46. Scan eBird Bar Charts 114
47. Visit birdcast.info during migration 116
48. Learn the sound of a bird near you 118
49. Tune in to social media 120
50. Take photos 122
51. View photos for habitat scenes and clues at Macaulay Library 124
52. Use eBird.org/explore 126

FINDING BIRDS WHILE DOING SOMETHING ELSE

53. Go to your local superstore 130
54. Bird from the backseat 132
55. Work outside 134
56. Visit faraway friends and family 136
57. Go for a walk in the rain 138
58. Take your kids to the playground 140
59. Play some golf 142
60. Do some yoga 144
61. Watch your step (on the beach) 146
62. Count vultures on a road trip 148
63. Scan the airport runways 150
64. Look for wild birds at the zoo 152
65. Visit a cemetery 154
66. Take the train 156
67. Take a bike ride 158
68. Go camping 160
69. Take a ferry ride 162

FINDING BIRDS THROUGH THE COMMUNITY

70. Join an Audubon Society chapter or bird club near you 166
71. Ask people with binoculars 168
72. Get your friend on the bird 170
73. Organize a bird walk for your community 172
74. Hire a bird guide 174
75. Get to know park or grounds staff 176
76. Participate in Global Big Day in May 178
77. Ask the locals 180

7 FINDING BIRDS ACTING CRAZY

78. Follow the caw of crows 184
79. Find fermented berries 186
80. Investigate any awkward flying or commotion 188
81. Visit a rookery 190
82. Stop for ducks dancing and pigeons pirouetting 194
83. Check bird deterrent spikes 196

8 FINDING BIRDS YOU'VE ALWAYS WANTED TO SEE

84. Wild Turkeys 200
85. Baby birds: Follow birds carrying something in their bill 202
86. Learn the scream of the Red-tailed Hawk 206
87. Hummingbirds: Visit red flower beds 208
88. See a Bald Eagle 210
89. Woodpeckers: Listen for drumming 212
90. Owls 214
91. See a stork 218
92. Pelicans 220
93. The fastest animal on the planet 222
94. A large pink bird 224

9 FINDING COOL BIRDS YOU MIGHT NOT HAVE HEARD OF

95. King of the pantry 228
96. North America's smallest falcon 230
97. Swifts: Hang out around chimneys in summer 232
98. Shrikes, aka "butcher birds" 234
99. Look around cows and horses 236
100. Sanderlings of the world's shorelines 238

10 ADVANCED BIRD-FINDING

101. Know your targets 242
102. Practice tracking birds in flight with common birds 244
103. Eke birds out of the sky / Put your binoculars to the test 246
104. Head out after a storm 248
105. Get a "little" lost 250
106. Look at night 252
107. Bird in the middle of the day 254
108. Stay in the car 256
109. Use the eyes in the back of your head / Detect flying shadows 258
110. Play the flock game 260
111. Prepare for takeoff to find a raptor 262

AS YOUR JOURNEY CONTINUES 265
BIBLIOGRAPHY 266
ACKNOWLEDGMENTS 267
ABOUT THE AUTHOR 268

INTRODUCTION

Are you ready to jump straight into an exciting nature documentary with birds wearing brilliant, ornate plumage, belting out impressive vocal repertoires, and doing funky dance moves to attract mates? You're in the right place. This book will open the door to the world of birds no matter where you are or what you are doing. Whether you're loading groceries into your car at the market, barbecuing with friends in your backyard or local park, or commuting to work by ferry, bus, or train, there are beauteous birds around waiting to be found. It may seem that the type of birds we see in nature documentaries are out of reach, living in remote, exotic locations. Some are, but, lucky for us, millions make the trek so we don't have to: Each spring, colorful warblers, vireos, cuckoos, wrens, orioles, and more fly up to the US from locations as far south as Mexico and South and Central America. Some sport shimmering, iridescent feathers and colors so vibrant they are almost indescribable; others may appear rather drab, but their wildly entertaining songs and behaviors more than make up for it. (And with a closer look, their plumage is just as intricate and beautiful.) Some of these migrating birds stay in the area to breed and raise young, while others make a pit stop to rest and fuel

Mourning Doves allopreening during courtship

up on caterpillar and spider snacks before moving on to their nesting destinations. (They'll pass through again in the fall as they make their way south.) This doesn't mean that spring and fall are the only seasons for bird-finding—it's a year-round event! And we're not just talking pigeons and starlings—resident mockingbirds, jays, hawks, falcons, and others always entertain, and winter visitors like ornate diving ducks and elegant loons mean you'll never be bored. Awesome, right? So how do we *find* them? There are more than a hundred ways. . . .

While you've likely encountered amazing stories or photos of birds in your area—fierce raptors, colorfully patterned songbirds, puffed-out "borbs," and awkwardly cute baby birds—it's easy to assume that these are rare, once-in-a-lifetime sightings. The truth is that some people are just really good at finding birds. Believe it or not, you *can* predict where a bird will be, what it will do next, and even when something insanely amazing is about to happen. There is no denying that a bit of luck is involved, but the bulk of bird-finding is wrapped up in a multitude of tidbits of experience, knowledge,

A secretive Wood Thrush migrates through Brooklyn, NY.

Young Barn Swallows in "Peeps" formation huddle together on a cold, rainy day.

A Hooded Merganser surfaces with a crustacean snack in New York City's Central Park.

and intuition gleaned from years of observing birds. That's exactly what you'll find in this book, and it will put you on the fast track to experiencing the bird show to the fullest—every day for the rest of your life.

Having spent many years "birding" (aka birdwatching), finding birds has become second nature. I can't wait to share my knowledge and experience with you on the pages that follow. I'll never be done with my exciting learning journey; with each new day and each new place I visit, I learn more and become even better at knowing where the birds are. You will too. Whether you regularly enjoy feathered visitors in your backyard or have never experienced a super-cool bird sighting, there are endless birds—and quirky bird behaviors— that await. So pick a "way"—any "way"—and see what unfolds. As you employ the ways to find more birds, you'll start to connect with the natural world on a whole new level and feel more a part of it. You'll want to reach out and touch the fuzzy branch of a staghorn sumac after watching a bird eating its fruit; you'll wonder what type of insect that flycatcher is eating as you literally hear the bird's bill snap shut. If this all sounds vaguely familiar, that's because it is—it's like being a kid again.

My goal is to make finding birds easy for you; there are no long hikes or special trips required on this journey. Most of the "ways" can be used right in your neighborhood and town. In fact, looking for

An American Robin snacks on staghorn sumac.

birds close to home—in a "patch"—is one of the best ways to find more. (See tip #18.) But just a warning—once you start tuning in to the birds around you, you might find yourself wanting to expand your territory. I never could have imagined I would fly to Colombia, the country with the highest bird biodiversity in the world, to look for birds.

Common Grackle fledgling found by its begging whine

While this book includes essentials for finding more birds (see "Bird-finding basics"), some of the best ways involve simply heightening your senses while doing something else ("Smell the scent of fallen flowers," "Beware of bark that moves") or positioning yourself so that the birds come to you ("Take a seat," "Work outside"). This book does not contain all possible ways to find birds. Since I want to make the experience comfortable and enjoyable for you, I have left out some things like going to a hawk watch (which may require a long trip and perfect winds), visiting a landfill (though I include "beautified dumps"), and taking an often seasick-ly pelagic boat trip (see tip #69 for fun alternatives). That said, any of these can often be a delightful bird-finding experience. As you apply the many tips in this book, you'll discover more of your own, all while becoming a veritable Sherlock Holmes of bird-finding.

Of course, once you start looking for birds, it's common to mistake all sorts of inanimate objects for avian creatures—whether it be the edge of a pipe jutting up from the top of a building, a piece of trash caught on top of a lamppost, or really anything that draws attention as not fitting into the expected landscape.

A Cooper's Hawk investigates a fake owl perched on a high-rise in Downtown Brooklyn.

You'll also have to contend with fake birds (most often fake owls), which are placed on rooftops, towers, and the like in an often-unsuccessful attempt to deter pigeons. (They're too smart for that!) Know that every birder on the planet has mistaken something for a bird way more than once, so laugh it off and cherish such stories for future telling.

It's important to note that the positive effects of finding birds will reach far beyond connecting with nature and will touch many areas of your life. Focusing on a bird will draw you into the present moment like never before and make you more mindful. Seeing the many struggles birds face will generate gratitude for the many conveniences we humans enjoy. You'll find yourself striking up conversations and making new friends while birding, and instantly helping and learning from each other. Putting a smile on peoples' faces will become a habit; when they ask what you're looking at, you might show them a Scarlet Tanager or Golden-crowned Kinglet and make their day. (And you could very well inspire them to start looking and listening themselves.) There are countless other things you'll notice; you might look forward to more walks in the rain or ferry rides in the winter. (I used to hate cold weather, but I now find myself traipsing through snow unfazed to watch Bald Eagles floating on blocks of ice.) Having to wait in a long line outside might become a welcome chore—you can look to the sky for soaring birds of prey or use the Merlin Bird ID app (see tip #45) to tune in to the area's bird sounds. While the show's ongoing script will be different for each of us, it will no doubt be life-changing.

I often wonder about all the amazing things I missed before birds were a part of my life. I don't want you to pass right by Barn Swallows feeding their young in midair or not notice the Red-tailed Hawk gazing down at you daily. Start experiencing the wildest free show on earth right now with *Find More Birds*. I hope you enjoy it!

A Barn Swallow delivers food to its young.

BIRD-FINDING BASICS

Common Grackle in Brooklyn, NY

1

FOLLOW SOUNDS

GREAT FOR:

- Finding migrating birds in spring when they sing to attract mates

- Discovering a bird's territory or planned nesting site, which they often defend through song

- Watching bird drama

SOUND IS THE SINGLE MOST IMPORTANT CLUE to finding more birds. While entire books have been written on the topic of how to identify birds by ear, it doesn't matter if you know what species of bird is singing, calling, crying, chattering, whistling, or screaming—you can still find it by following its sound. During a class I teach at the Brooklyn Botanic Garden, we were practicing the topic of the week—detecting and following bird sounds. A raspy and repetitive chatter in the distance led us to a feather snowstorm emanating from a large oak. As the plumes landed in our hair, a wildlife documentary unfolded right in front of us—a Red-tailed Hawk was perched above, clutching its pigeon prey. As if that wasn't exciting enough, half a dozen Common Grackles were mobbing the hawk, jumping on it while making the noises that had led us straight to the show. To find more birds—and more bird drama—always follow sounds.

The more you do this, the more you'll start to notice yourself organically learning and recognizing the sounds of bird species. Kick-start this process by listening closely to the qualities of the sounds you hear—rhythm, tempo, timbre, pitch—and approach it as a form of music appreciation. Before you know it, you'll be able to pick out the cheerful tune of a Song Sparrow or the piercing call of a Blue Jay. Actively listening to birds will also improve your skills in detecting distant birds and in associating certain sounds with bird

Red-tailed Hawk

behaviors. (Birds often make "alarm calls" to warn others of a nearby predator and, as with the previously mentioned grackles, "mobbing calls" to recruit help during an attack on a predator.)

GOOD TO KNOW: *Listen to the songs and calls of birds near you with the free Merlin Bird ID app (see tip #45); it will even identify bird sounds for you! (And if you are hard of hearing, Merlin can do the listening.) A great way to memorize a bird's song is to* watch *it sing, whether out in the field or with videos you can find on YouTube or at macaulaylibrary.org.*

Warblers such as this Canada Warbler are tiny and tough to spot. Look for movement in trees and shrubs during migration to find them.

2

LOOK FOR MOVEMENT

A CLOSE SECOND to following sounds, looking for movement is an essential tip for finding more birds. Granted, this can be difficult on a windy day or when leaves are falling in autumn, but tuning this skill on tranquil-weather days will turn you into a bird-finding pro. Try this exercise: Pick a place to sit or stand for fifteen minutes. Scan the landscape and notice what is moving—it could be planes, dragonflies, a squirrel, or a possum, but hopefully also birds. All the things you notice moving will help you develop better detection of movement. Using a soft gaze helps; try not to focus on any one thing as you scan the area and you'll pick up on slight movements more easily. When you notice a bird, change to hard focus and lock on the thing as if you were a pro tennis player about to return a 150 mph (241 km/h) serve. Watch the bird in motion—whether it's flying, running, hopping, or wading—and observe it and enjoy. If you have binoculars (see tip #7), keep your eyes on the bird while

Much like the Black Phoebe, a flycatcher of the western US, Eastern Phoebes sally out from low perches to hunt insects. To find them, look for movement along railings, posts, and trees adjacent to buggy lawns and meadows.

you raise them to your eyes. (Maintaining a view of the bird and/or source of movement from naked eyes to binoculars is not easy, but it's worth taking the time to develop this skill.)

Use the same soft-gaze scanning technique on a single tree, a tiny shrub, or a patch of leaf litter. During migration, this is a great way to find warblers, which are tiny and difficult to see; stay a while and look for subtle movement as these birds hop and flit through the habitat, foraging for insects and other arthropods. Soon, you'll be able to detect birds in their favorite hiding places; you'll notice how the gentle tug of a tree branch that snaps quickly back into position often signals a bird, while a stronger, longer pull or drag of a branch is usually a squirrel.

GREAT FOR:

- Migrating warblers and kinglets that move frequently as they forage through trees and shrubs
- Flycatchers that repeatedly sally out from a perch to snatch an insect
- Hard-to-see birds whose plumage blends in with their habitat

American Robins in survival mode,
waiting for ice to melt after a snowstorm

3

CHECK FRESHWATER SOURCES, EVEN DRIPS

GREAT FOR:

Clear and longer
views of birds as
they drink from
tiny puddles and
splash in shallow
baths

DRIP, DRIP, DRIP. The leaky spigot in the campground
may seem wasteful, but the local crows are certainly
putting it to good use. It's their drinking fountain. I
discovered this when we unknowingly pitched our tent
near the spigot and were awakened each morning by
caw-caw-caw-ing as a trio of crows flew in to fight over
the drips. (It's just as well they woke us up early, as we
had a lot of birds to find.)

Most birds need water for drinking and bathing, so
knowing the local watering holes is a great way to find
them. Shallow, gently sloped edges of ponds are ideal
places to look, especially those surrounded by trees or
other perches where birds can preen after a bath. Water

that moves or drips also catches birds' attention, so be on the lookout for streams or active fountains. Surfaces of ponds and lakes are frequented by swallows skimming a drink in flight; swifts and many nightjars hydrate the same way.

White-throated Sparrow

Some of the best places to find birds enjoying water are temporary. Small, shallow puddles that form after a rainfall or sprinkler session are perfect for songbirds. Check for these impromptu bird baths on the ground, on large rocks, or on anything else where 1 to 2 inches (2.5–5 cm) of water might collect. I once found a group of twelve young Cedar Waxwings bathing on a lawn, splashing around in mini pools of water that had formed between clumps of grass.

When their regular water sources freeze, birds often seek out leaks, drips, and melting ice. After a heavy Brooklyn snowfall, when every pond and puddle was frozen solid, I spotted American Robins lining a wooden bench patiently waiting for ice to melt and drip down the bench backrest; nearby, a Gray Catbird balanced on a slippery banister to lick off droplets of water.

GOOD TO KNOW: *Not all birds need fresh water to drink. Some can stay hydrated through their food. Desert-dwelling Black-throated Sparrows, for example, obtain water from their diet of juicy insects and succulents. Others, such as gulls and albatross, have salt glands that allow them to drink seawater and expel the salt.*

4

STAY IN ONE SPOT FOR AT LEAST 10 MINUTES

GREAT FOR:

- Witnessing behaviors like courtship, mating, and territorial defense

- Seeing a spot come alive with more and more birds as they get used to your presence

- Maximizing your chance of finding a rare bird or a fleeting flock

ONE COLD NOVEMBER DAY I came upon a dozen hawthorn trees packed with gleaming red berries. Yay! Great habitat! Birds love berries, right? (See tip #31.) At first, I didn't see a single bird, but after a few minutes of scanning the rubyscape, my eyes adjusted, and several American Robins that were feasting on the fruits came into view. With each passing moment, I spotted more birds until dozens brought the trees to life. As I watched them juggling berries on their tongues, the high-pitched whining of Cedar Waxwings somehow broke through the noisy chatter of the robins. As I looked closer, I spotted several of them among the densely fruited hawthorn branches.

As you look for birds, you'll soon realize that the landscape before you may appear devoid of bird life one minute, then come alive with birds the next. It could be that the birds are there and you just need a few minutes for your eyes to adjust and detect movement. Or maybe there really are no birds, but a flock or hawk could fly in any minute. If you come across a promising habitat—a stand of trees or dense shrubs, marsh, mudflats, a tree-lined pond—give it a wait. Ten minutes is perfect, and you don't have to actively look the entire time. Birds seem to like it when humans appear uninterested, and they often go about their foraging unperturbed from just a few feet away. So do whatever you like while you wait—sit down, read a book, submit an eBird list, check your phone. As you're hanging out, be aware of any sounds, or of movement catching the corner of your eye, and scan periodically for increased activity.

American Robin

Golden-crowned Kinglet

5

LOOK DURING MIGRATION

THERE MAY BE HUNDREDS OF SPECIES on the list of birds seen in your local park on eBird.org (see tip #46). How is this possible? While there are plenty of resident birds, many more stop by to rest and refuel during their migration journeys. The list includes also rarities, some of which have made only a single appearance. To maximize the number of birds you find, you'll want to be out and about in your yard or local park—or watching from your window—during migration.

So, when exactly is migration in the US? The short answer is that the best times are May in the spring and September and October in the fall. But spring songbird migration actually starts in mid-March and continues into June, and fall migration begins in June and continues into November or even December. That makes it sound like migration lasts nearly a year, right? Well, it kind of does, because birds are always on the move somewhere. (If you factor in ducks and other waterfowl that migrate in winter, you have a yearlong show.) The important thing is to be on the lookout in May, September, and October to find the most birds.

As you can see, spring migration is shorter (including its peak). At this time, birds are in a hurry to get to their nesting locations, where they have to establish territories, attract mates, build nests, incubate eggs, and raise their young. (It's a lot of work, especially to pack into a couple of months.) So they don't stay long when passing through—they fuel up, rest, have a drink, and are usually up, up, and away the next day. In the fall, migrating birds stick around longer, since the grand task of breeding is complete. They can take their time as they head south to places like the Caribbean, South America, and the southern US. This means that peak fall migration is much longer, and visits by different species are more staggered, so September and at least into mid-October will be rockin'.

GOOD TO KNOW: *See tip #47 for how to get bird migration forecasts!*

GREAT FOR:

- Finding many different bird species at once—some passing through, others staying to breed

- Seeing vibrant, colorful plumage of migrating birds in spring, especially warblers and tanagers

GET A BIRD FIELD GUIDE . . . MADE OF PAPER

GREAT FOR:

- Getting to know the many bird species you can find in your area and beyond

- Internalizing the shapes and sizes of various bird families

- Studying birds' "field marks," especially to learn how to differentiate similar-looking birds like sparrows and gulls

- Taking on trips (especially a field guide for a specific region)

"YOU CAN NEVER HAVE TOO MANY FIELD GUIDES." These words of Lucy Duncan, my mentor, have always stuck with me. Maybe that's because they allow me to continually make guilt-free purchases—a guide on warblers, another on gulls, one on the birds of Texas, yet another on birds of Puerto Rico. Then there's my latest fascination with field guides in Spanish. In all, I have over thirty bird field guides.

It may seem like the main purpose of a field guide is to help you identify birds while out in the field, but don't be fooled. Looking through one at home will help you find more birds. As you thumb through your guide, admiring many birds, you'll be subconsciously uploading the illustrations and photos into your memory bank. And since many guides are organized by loose bird family groupings, such as sparrows, hawks, and herons, you'll also be associating birds of the same size and shape together in your brain. Many also have information on each bird's preferred habitat, behavior, even flight style—a host of bird-finding clues! All of this input means that when you're out and about, you're more likely to notice an interesting silhouette that matches something in your field guide, or to distinguish a bird with a distinctive characteristic—especially something like a longer tail or a uniquely shaped beak—from the pack.

Since many field guides are inherently intriguing (and beautiful), if you keep one in plain sight on your coffee table or desk, you'll be more likely to thumb through it often. Start with one that covers birds in your area. But be warned—you may soon find that you can never have too many field guides.

Peterson *Field Guide* Roger Tory Peterson *of Eastern and Central North America* HMH

BIRDER'S GUIDE TO THE RIO GRANDE VALLEY LOCKWOOD McKINNEY PATTEN ZIMMER ABA

A GUIDE TO THE BIRDS OF PUERTO RICO AND THE VIRGIN ISLANDS RAFFAELE

NATIONAL GEOGRAPHIC FIELD GUIDE TO THE **Birds** of NORTH AMERICA Dunn Alderfer

GUÍA PARA IDENTIFICAR AVES POR SU COMPORTAMIENTO

STOKES FIELD GUIDE TO **BIRDS** WESTERN REGION

Field Guide BIRDS COLOMBIA MILES McMULLAN R-N

BIRDS OF NORTH AMERICA Robbins Bruun Zim Singer GOLDEN

Raffaele *Birds of the West Indies*

KAUFMAN Field Guide to **Advanced Birding** HMH

BIRDS OF SOUTH AMERICA ERIKE NADA RIBEIRA

ROSEMARY MOSCO A POCKET GUIDE TO PIGEON WATCHING workman

BIRDS of **TEXAS** Arnold & Kennedy

KAUFMAN Guía de campo *a las* aves de Norteamérica ASSO HMH

A FIELD GUIDE TO SEABIRDS OF THE WORLD Peter Harrison ISBN 0-XXX-XXX-X

NATIONAL AUDUBON SOCIETY THE **SIBLEY** GUIDE TO **BIRDS** Knopf

Bird FIELD GUIDE TO **BIRDS** OF NORTH AMERICA Brinkl

Robert S. Ridgely and P.J. Greenfield The Birds of Ecuador FIELD GUIDE

Stephenson and Whittle The **Warbler** Guide

Up close and personal with a Eurasian Jackdaw at Richmond Park in London, UK

7

GET BIRD-WORTHY BINOCULARS

IF YOU DON'T HAVE BINOCULARS, get a pair as soon as you can so you don't miss out on a thing. You'll be able to enjoy all the birds you find to the fullest, with close-up views of gorgeous feather patterns, super-sharp talons, fascinating bird behaviors, and more. Believe it or not, you don't have to spend a bundle to get binoculars that are great for birding. It's not the brand or the cost that matters most—it's the specifications. You could buy a $2,000 pair of binoculars with the wrong specs for birdwatching, and the person next to you with a $40 pair will be having a much easier and more enjoyable time observing birds.

Binoculars with one of the following specs will make it easy to get birds into view and keep them there.

- 8×42mm (arguably the most popular)
- 10×42mm
- 7×35mm

The first number is the degree of magnification, and the second is the diameter in millimeters of the objective lens (at the far end of the binoculars). A large objective lens size means heavier binoculars, so stick with the recommended specs for comfort and safety. Note: For a compact option that you can take anywhere, 8×25mm is a popular choice.

To find more birds, the 7× or 8× magnification is surprisingly better; in most brands of binoculars, the 8×42mm (or 7×35mm) model will allow you to see a larger area through the lenses, called "field of view" in binocular lingo. So, if you're observing a bird through binoculars and another lands nearby, or a large flock passes a bit off to the left, you might see it with 8× but not with 10×. The 10×42mm are also more subject to hand shake, making it more difficult to hold steady on a bird. All of that said, I have used all three and have been pleased with every one of them. I can currently be found sporting a pair of 8×42mm.

No matter what model you get, you'll need to set up the focus on first use, otherwise you'll always be wondering why the birds are blurry! Focus on the same object (a street or park sign is a good choice) one eye at a time: Close your right eye and use the main focus knob for the left eye. Then close your left eye and use the diopter adjustment for your open right eye. (The diopter adjustment is usually a dial on the right eyepiece; check your manual for exact location and details.)

GREAT FOR:

- Unbelievable views of birds, their detailed plumage, and what they are eating

- Identifying birds after zooming in on and analyzing their field marks, habitats, and foraging styles

- A front-row seat to entertaining bird behaviors, including courtship and mating rituals

GOOD TO KNOW: *As of this writing, you can purchase a pair of Bushnell Falcon 7×35mm binoculars for under $35! I recommend this pair to all of my students and bird walkers, and they all love them. Though some have informed me they have experienced "binocular shaming"; other birders have told them they need to get a better pair! Feel free to laugh off such comments while saving hundreds—or even thousands—of dollars and enjoying birds through your bird-worthy binoculars.*

SCAN THE SKY

GREAT FOR:

- Large, V-shaped flocks of pelicans, cormorants, ibis, geese, and cranes
- High-soaring hawks, eagles, and vultures

SPOTTING HIGH FLYERS, even in the distance, is easier when you're at their level. You'll often notice them from a mountain or a bluff, but from down below, they are easy to miss. Get in the habit of scanning the sky periodically, especially in early to mid-afternoon, when hawks and vultures ride hot air thermals that rise up from the earth, allowing them to soar effortlessly without the flap of a wing. Looking up will also increase your chances of spotting large V-shaped flocks of geese, cormorants, pelicans, and ibis, or irregularly shaped flocks of songbirds. Mixed in with a bit of luck, you never know what you might find. One evening while walking in downtown Pensacola, Florida, I noticed a super-wide V formation approaching in the distance; I watched as the rare flock of forty American White Pelicans passed over the traffic lights.

Stopping and chatting is also a surprisingly great way to find soaring birds; your peripheral vision will kick in and automatically detect birds in flight. While in Portland, a couple stopped to ask me about the many Cedar Waxwings flycatching from a nearby snag (a standing dead tree). As we talked and watched the birds' aerial acrobatics, we were set up perfectly to spot dozens of American White Pelicans passing across the sky! The first Bald Eagle I ever found in my birding patch was also a "stop and chat" find. It was sheer luck that my friend Bob was standing a couple of stairs up as we talked, affording me the perfect angle to notice the eagle soaring high in the distant sky.

American White Pelican flock

TAKE A SEAT

GREAT FOR:

- Close-up, even eye-level, views of birds

- Drawing birds out of nearby hidden locations (they'll be more comfortable with your presence)

IMAGINE WHAT WE LOOK LIKE to a small songbird. We're giants! So it's no surprise that birds often flee as we approach. The good news is that many birds that take off at the sight of us will return if we simply sit down. Why? Reducing our height by half seems to make birds feel much less threatened. And the birds like that spot where you initially saw them—they were there for a reason—probably to partake of some tasty insect, seed, or berry snack.

So take a seat and wait for the show. You can find more birds the lower you sit; a bench or a log works well, but sitting on the ground is even better. My closest encounters with birds have happened while seated on lawns, including having a Golden-crowned Kinglet hop onto my knee and give me a sweet look. This past fall migration, I noticed tiny warblers flying around a small pond in my patch. I wasn't sure what they were, but took this as a cue to sit down close to the pond to find birds. Within five minutes, I was surrounded by feathered gems. A Magnolia and Cape May Warbler hopped on the lawn less than a foot in front of me. To my right, beneath the shrubs lining the pond, the secretive Lincoln's Sparrow peeked out and started to forage, while an Ovenbird strutted out to its right.

Sitting down works especially well in an area where you see several birds. Since they often hang out in groups, once you sit down, you'll notice even more.

GOOD TO KNOW: *It's always worth getting a little mud on my jeans to find more birds, but carrying a thin bag or cloth to sit on can be handy, especially during migration, when you're more likely to suddenly encounter pockets of bird activity.*

A secretive Lincoln's Sparrow on lawn's edge

10

AVOID SUDDEN MOVEMENTS AND REMAIN CALM

GREAT FOR:

- Preventing birds from fleeing, hiding, or stopping what they are doing

- Closer, unobstructed views of birds

I GET VERY EXCITED about birds. When I first started looking for them, each time I saw a new species I was probably flailing my arms in joy and pointing to the bird over and over. That was until my mentor, Lucy, told me that pointing is a major arm movement and can startle birds. Same goes for sudden turning, raising your binoculars too quickly, or jumping up and down when you get a "life" bird (a bird you have never seen before). Clothing or backpack straps waving in the wind are distracting as well. No movement is ideal. Secure all your loose ends!

Along the same lines, it helps to stay calm as you look for birds, or at least appear that way (as in my case). Let the birds think you don't care about finding them. As you're walking, if you notice a bird perched close in front of you, it's best not to stop suddenly or meet eyes with it. Keep moving and pass it by; when you're about 15 feet (4.5 m) ahead, turn back to observe it. Similarly, when you hear a bird, turn or move slowly toward the sound. When I heard a bright chirp directly behind me, I used my best slow-motion movie scene imitation to find this Orange-crowned Warbler among the winter blooms of the common yarrow.

Orange-crowned Warbler

11

LOOK FOR BIRDLIKE SHAPES AND SILHOUETTES

GREAT FOR:

- Raptors perched on high points with the sky as a background (treetops, light posts, buildings)
- Power line perchers like flycatchers, falcons, swallows, and shrikes
- Picking out lanky, long-necked birds like herons and egrets
- Birds near water at night (when shapes contrast well with light reflecting on the water)

WHILE IT'S TRUE that some birds' silhouettes are much easier to detect than others—the large Great Blue Heron against a beach sunset or a soaring hawk back-lit against the afternoon sky, for example—once you start looking, you'll quickly learn to find even small birds by their silhouette. The best way to start honing this skill is by checking open perches—on top of street signs, buildings, light posts, towers, cables, treetops, and the like. Scan leafless trees in winter and early spring, at which time you will also encounter old birds' nests—a clue to check the same trees for active nests come spring.

Once you find a bird silhouette, snap a photo so you can later analyze the shape for identification clues. Then, if possible, move to a place where you can observe the bird with your back to the sun; this will afford you a better view with more colors, details, and ease of observation. If you are duped by a supposed bird silhouette, you're in good company; this happens to birders all the time. Fake plastic owls and hawk kites are obvious tricksters, but I've been fooled many times by a fixture on a light post and other things that have caught my attention as looking out of place.

The small silhouette of a Belted Kingfisher stands out when scanning the Manhattan skyline.

Great Blue Heron

FIND THE HIGH POINT

GREAT FOR:

- Birds foraging or roosting high in the tree canopy
- Raptors, flying or sitting on one of many perches better visible from high ground
- Getting an overview of habitats available in a large park or preserve (scan for ponds, marsh, prairie grass, red flower patches, etc.)

WHETHER YOU'RE IN a small park or a large wildlife refuge—really, any landscape that isn't flat—find the highest point. This will give you a wide, bird-finding view of the area so you can more easily scan the treetops and landscape for movement and discover pockets of habitat worthy of further investigation. Finding soaring raptors also becomes much easier with an unobstructed view of the sky and less neck craning. Many refuges have hawk viewing platforms for this very purpose; in forests, such structures are often above the tree canopy, affording a view of birds that forage and perch in the treetops.

If there's a tower or castle to climb, go for it. You might be so high that you can find hawks flying at eye level or below. On a trip to Puerto Rico's El Yunque rain forest, we hiked up to the Mount Britton Tower and enjoyed stunning views of a trio of Red-tailed Hawks flying just beneath us and over the Caribbean landscape.

Sometimes being up a level gives you the perspective you need to find a bird. In Colombia, we heard a bird drumming (see tip #89) above us as we enjoyed a close-up view of tanagers at feeders down below. No matter how often we looked, we could not see a woodpecker anywhere. A couple of hours later, we walked up to the roof of the *finca* (estate) to take in the view and laughed as we encountered the noisemaker in a tree at eye level: a Lineated Woodpecker.

Lineated Woodpecker in Colombia

DON'T LOSE HOPE IF IT FLIES AWAY

GREAT FOR:

- A second chance to get great views of a bird
- More quality time with birds
- Learning and predicting bird behavior

IT CAN BE FRUSTRATING when you encounter a bird that takes off so fast you have no chance to observe it or to figure out exactly *what* you've found. When that happens, don't lose hope. Time and time again I see people throw in the towel the second a bird takes flight, thinking they've lost their chance at more quality time with a bird. Instead, practice reacting to a bird taking off by locking your eyes on it and following it as it flies. The bird will often land nearby. If not, unless you see the bird fly off into the distance, there's a good chance it's still around.

And now you can employ tip #10, moving calmly and quietly as you try to relocate the bird, bettering your chances that it won't flee when you find it. One tactic you can use is to move away from the area from which the bird flew and eye it from afar. Birds often return to the same location after they are "flushed" (scared away) once people and other perceived predators move on.

The more time you spend observing birds, the more you'll be able to predict their behavior, which includes what they might do when flushed. You'll be able to think like a bird. I was caught off guard when a raptor suddenly appeared from behind a berm and flew off behind me. I could tell it was carrying a pigeon in its talons, which meant it would be heading somewhere to feast on its prey. My best guess was that it just made a short hop and stuck to the berm. I walked to the back of the berm adjacent to the Brooklyn-Queens Expressway to find it: a Cooper's Hawk plucking feathers from its prey.

Cooper's Hawk preps its meal on a berm.

14

CHANGE YOUR ROUTE

GREAT FOR:

- Discovering new bird habitats near you
- Increasing your chances of finding a rare bird
- Finding nests

ONE THING IS FOR CERTAIN—there's just too much ground to cover to find each and every bird that surrounds you. But you can maximize the number of birds you encounter by changing your route every so often. Even though passing the same habitats day after day is one of the best ways to improve your skills—and quickly turns you into a local bird-finding expert—you could be missing out on some really cool birds just around a different corner.

On a birding trip in Colombia, our car overheated and we had to wait a few hours to get alternative transportation. That's not such a big deal when you're with a bird guide in the country with the largest number of bird species in the world. My guide and I decided to check out a dirt road that was not along his usual tour route. Within a few minutes, we started to hear strange squawking noises and followed the crescendo of sounds to find a breeding colony of Black-crowned Night-Herons. Closer to home, when a new native prairie grass habitat opened in Brooklyn Bridge Park, I checked it a few times but stopped visiting after not having seen many birds. A year later, I decided to give it another try and found a super-rare Painted Bunting.

Try a different route on your next walk and see what you find!

A rare Painted Bunting in Brooklyn

15

STAY FUELED—ALWAYS HAVE SNACKS ON HAND

GREAT FOR:

- Staying safe and hydrated
- Keeping your bird-finding senses sharp
- When you want to stay longer to find more birds!

I LEARNED LONG AGO that it's better to have more than enough food on hand when looking for birds. In search of the near-threatened Red-cockaded Woodpecker in Tate's Hell State Forest in Florida, my birding buddy Connor and I wound up getting lost. On top of that, we had left all our snacks—and water—in the car. My adrenaline rush from finding the woodpecker took a nosedive when we couldn't find our way back. Heading toward the forest edge to look for a road was fruitless; each dirt road went on for miles with no sign of a car. We retraced our steps at least four times under the strong afternoon sun until we finally saw the parking lot in the distance.

Bringing snacks and water is not only safer, it will also help you find more birds. You can stay out in habitats longer and enjoy a picnic in one of the many beautiful spots you'll encounter. Sitting down for a bite and a rest also activates tip #4 ("Stay in one spot for at least 10 minutes").

On a recent trip, my backpack was filled with protein-rich snacks: spicy dried chickpeas, dried green bean pods, cheese sticks, vacuum-packed tuna, and more. I didn't eat even half of them, but just knowing I was prepared meant I could continue to explore . . . and spend extra time observing cute tern and skimmer chicks!

GOOD TO KNOW: *While it's tempting to share food with birds we meet, it's best to let them find their own. Scattering snacks can attract hordes of pigeons and gulls and make birds (and squirrels!) associate humans with a food source. Bread and bagels can make birds feel full and stop them from eating something more nutritious that they need to stay on top of their game.*

Gray Catbird with mantis

16 DRESS FOR SUCCESS

GREAT FOR:

Blending in with the landscape to avoid startling birds

WHEN I STARTED BIRDING while living in Florida, there was one thing about the pastime that scared me to no end—ticks. My friend Connor and I cooked up a plan so we could easily spot any of these critters hitching a ride. We arrived at the beautiful Naval Live Oaks on the Gulf Islands National Seashore, outfitted in bright white sweatsuits, determined to keep our skin tick-free, and ready to spend quality time with birds we'd seen there before: the Pileated Woodpecker, Brown-headed Nuthatch, Tufted Titmouse, Northern Flicker, and more. It didn't take long to realize that the birds could spot us from a mile away in our bright white garb and had absconded well before we approached; after over an hour of not seeing a single bird, we headed back to the parking lot and picked a handful of ticks from our sweatsuits. We never wore them again.

It's not that birds can't see you approaching in clothes of any color, but wearing earth tones and blending in with the landscape is your best bet when trying to find birds. White is not a dominant color in most habitats, so it shines like a beacon, warning birds of your arrival; they'll pick up on it quickly and flee or hide.

GOOD TO KNOW: *There are plenty of natural insect and tick repellents (ticks are arachnids, yikes!) to choose from. Ticks also stay away from the scent of essential oils such as lemon, orange, and peppermint.*

Take a cue from this House Wren and wear earth tones to blend in with the surrounding habitat.

A different Groove-billed Ani, not disturbed by playback

17

AVOID STRESSING BIRDS OUT— DON'T PLAY RECORDINGS

GREAT FOR:

- Ethical bird-finding that respects birds and their habitats
- Letting birds forage or rest without wasting energy investigating fake threats
- Preventing birds from fleeing the area to avoid territorial competition

IN THESE DAYS OF SMARTPHONES and apps that contain birdsong, one might think it's a no-brainer to use sound recordings as a tool to attract and find more birds. But according to the Code of Birding Ethics of the American Birding Association (ABA), which promotes respect for birds and their environment, we should "limit the use of recordings and other audio methods of attracting birds." While some still use this "playback" to attract birds, it can have the opposite effect. I recall watching a rare Red-headed Woodpecker happily pecking away on a snag in Central Park until someone leading a bird walk came by and started playing a recording of its call. This attempt to draw the bird in closer for crowd-pleasing photos backfired; the bird took off and out of sight. Much to everyone's delight, a few minutes after the playback stopped, the colorful woodpecker returned to forage in peace.

When I started birding, I used playback a few times because I saw others doing it, but I quickly stopped after I successfully drew a migrating Groove-billed Ani out of hiding. It hopped around looking for the source of the sound—maybe a potential mate, competitor, foraging friend—and found none. I felt like I had just played a cruel joke on the bird. At that point I realized that birds have more important things to do than be duped by me; their survival depends on fueling their bodies during migration, conserving their energy, and staying safe from predators. Why was I acting against the best interest of these creatures I loved?

While the use of playback is important for scientific studies and monitoring that contribute to bird conservation, it's tough to argue that using it solely to get a look at a bird isn't disrespectful deceit. While some experts say that playing recordings doesn't have a negative effect on birds outside their breeding grounds, it does distract them from what they are doing—often foraging or eating. (I wouldn't want someone calling out to me every few minutes while I was trying to finish my dinner.) Plus, birds sing to attract a mate and defend territory. Hearing another of the same species may put birds into defense mode and drive them away from a potential nesting site. Using your bird-finding skills and remembering tip #10, "Avoid sudden movements and remain calm," are much better ways to find birds and prevent them from fleeing the area.

GOOD TO KNOW: *No playback was used to obtain any of the photos in this book.*

18

PICK A PATCH

GREAT FOR:

- Honing your bird-finding skills
- Spending quality time in nature
- Witnessing fascinating bird behaviors, including courtship, mating, nest-building, territorial disputes, and more
- Reducing your carbon footprint through "green" birding (walking or biking) close to home

THE MORE TIME YOU SPEND looking for birds, the more you'll find. But who has the time? Patch birders. A "patch" is a place close to home or work where you go to find and observe birds. Since it's close, you'll be able to visit often, even when you're short on time. Your patch can be your backyard, your local park, a group of shrubs at the end of your block, a pond adjacent to a parking lot at work, even a tree outside your window. Right now, mine is Brooklyn Bridge Park in New York City, where I go to find birds most days of the week, whether for fifteen minutes or several hours. (I have found 182 species there so far!)

By picking a patch, you'll gain deep knowledge of how, where, and when to find birds there. It will also become your field laboratory where you can fine-tune your birding skills whenever you like. For example, birding by ear will quickly become easy in your patch. And you'll get to know its habitats and features so well that you'll instantly notice something different there—and find more birds.

One of the best things about patch birding is that, in addition to birds, you'll find super-cool plants, insects, and other creatures and become an expert on how the elements of your patch's ecosystem depend on each other and function together. It's a nature show unfolding right outside your door; you'll want to be there to see what happens next.

GOOD TO KNOW: *Birders often choose a patch that is "under-birded" where no one else is regularly looking for birds. It's super exciting to be able to fill in gaps in citizen science data by submitting your patch sightings to eBird.org. Before long, others will be showing up to get in on all the bird action you've found!*

Hermit Thrush on redbud

Common Tody-Flycatcher

19

KEEP A LIST

GREAT FOR:

- Motivation to get one more bird for your list, and another, and another. Finding birds becomes an exciting collection game.

- Contributing to bird conservation by submitting your lists to eBird

- Having a record of the many cool places you've visited on your bird-finding adventures!

THERE ARE MANY TYPES of bird lists—life lists, year lists, state lists, county lists, patch lists, even wedding lists— with the number of bird species spotted during a time and/or at a location. The most common is a life list, consisting of all bird species you have ever seen anywhere. The entertaining pursuit of a year list is even portrayed in a movie, *The Big Year*, based on a book of the same name. "Listing," as it's called in the world of birding, can be intense or casual, and it's certainly not a prerequisite for watching birds.

On the other hand, keeping a list can help you find more birds. It's a collection game—once you add one species, you'll want to add another, and another. One year, as part of the Brooklyn Birdathon (an event to raise money for the installation of bird-safe glass), our team had spotted forty-nine bird species in Brooklyn Bridge Park; we were done. But as we left the park, finding a fiftieth species for the list was just too tempting. Desperate and determined,

I suggested we spend five minutes scanning the surrounding sky for raptors. How could anyone refuse? Using the reach of our binoculars (see tip #103), we soon spotted a faraway bird soaring like a hawk, too distant to identify. Within minutes, an Osprey came into view, clocking in at #50.

Osprey making the list for a Birdathon fundraiser

Adding birds to your life list can be extremely exciting, especially when you have your eyes set on the prize in advance. Before traveling, choose your target species for the trip and use eBird.org/explore (see tip #52) for the best chance to find them. When traveling to Colombia, my main target was a Common Tody-Flycatcher; my bird guide led me to its habitat, and I watched in awe as it devoured a large moth.

Start by making a list of birds you see each time you're in your backyard, local park, etc. If you're not sure what bird you've found, get help from the Merlin Bird ID app (see tip #45) or note the characteristics of the bird and investigate later. You can use the eBird app or website to record your lists, which makes it easy to view, filter, and sort them later. Plus, you'll be contributing to a citizen science project that makes data freely available to all, including researchers and conservationists working to preserve birds and their habitats.

GOOD TO KNOW: *eBird lists are a great way to review all the cool places you've visited, whether you saw a lot of birds there or not! Even if you only find a single bird, submit a list. Then, if you're racking your brain to remember how to find that cute neighborhood or great restaurant you visited a while back, all you have to do is scan your eBird lists for the general date and location.*

FINDING BIRDS

at their

FAVORITE RESTAURANTS

Henslow's Sparrow

20

DISCOVER HIDDEN GEMS IN LEAF LITTER

ON BIRD WALKS DURING MIGRATION, a reliable bird-finding trick is to stare at a patch of leaf litter for at least ten minutes. We do this to find an Ovenbird, a migrating wood warbler that constantly struts through the leaves like a mini chicken. As we watch and wait, the Ovenbird suddenly enters the scene. The sight of it gobbling down insect after insect is so entertaining we don't want to leave.

Birdwatching evokes images of people looking up, scanning the sky and treetops for things that fly. But birds can be found all around, often right at our feet, undetected. Even at close range, they have a great place to hide—in the leaf litter under trees and shrubs. Here, birds such as sparrows and thrushes forage for insects, spiders, and worms hidden beneath the leaves. Often there are dozens of birds before you, but they're so well camouflaged with their earth-toned plumage that you might not see a thing. Lucky for us, leaf litter is noisy. As thrushes and grackles flip over leaves to find a tasty treat, and sparrows "double-hop scratch" to dig for dinner, they can't help but give us a clue as to their whereabouts. When you encounter leaf litter under trees and shrubs, stop and listen as you slowly scan the leaves for movement. Also scan leaf litter on lawns for subtle movement; the rare Henslow's Sparrow pictured on the previous page didn't seem to be making a sound as it devoured arthropods on a cemetery lawn.

Winter is a great time to find birds in the underbrush. The stillness of the season accentuates the sounds of birds foraging; leaf litter comes alive as dozens of sparrows displace its layers in search of food. During the colder months, look for White-throated Sparrows in the eastern US and White-crowned Sparrows in the West. In much of the South, Hermit Thrushes can be found hopping around as they overturn leaves. In the Southeast, you'll hear more constant crunching as Common Grackles walk rather than hop through the dead leaves.

GREAT FOR:

- Sparrows hopping and scratching in the leaves
- Migrating and wintering thrushes overturning leaves with their bills
- Ovenbirds strutting through patches of leaf litter during migration

Common Goldeneye

21

SCAN THE WATER . . . TWICE OR THRICE

GREAT FOR:

- Diving ducks that pop up suddenly to the surface
- Loons and grebes
- Distant waterbirds, including large, far-off flocks of waterfowl

EVEN THOUGH WATER is one of the easiest places to find birds, it's also a habitat where you can easily miss a bird at first glance. Some ducks, loons, and grebes can stay underwater for anywhere from thirty seconds to more than a minute. If you keep scanning the water, a pop-up surprise—like a Bufflehead duck with a white and iridescent head or a funky-looking Pied-billed Grebe—may surface.

Even above water, it can be tough to detect a different duck. This past fall, I saw the usual waterfowl suspects at Brooklyn Bridge Park's mini beach. One was sleeping on a far-off granite block, its bill and head tucked into its body feathers; it looked a bit like a sleeping Brant (a type of goose), but something felt different about this bird.

Luckily, I scanned the area again at just the right moment, when it was descending into the water and I could see its distinctive features. It was a Ring-necked Duck, and a new species for my birding patch.

If you're looking for birds in the water among boats, piers, docks, bridges, and jetties, birds can move in and out of view quickly, so finding more birds requires repetitive scans. Plan to spend some time in such a habitat watching for birds surfacing, flying in, and appearing from hidden places.

Bufflehead

If you come across a small- to medium-size wetland packed with ducks, use your field guide to identify all of the male ducks you can, and study the subtle features of their female counterparts. (Many female ducks look similar and can be difficult to find in a field guide.) Keep scanning for something different in the crowd. Finding more species among hundreds of waterfowl is a fun game, especially since they are large and easy to observe and study.

Along the coastline and in larger wetlands, ducks may be very distant and require a spotting scope to differentiate. Focusing on movements and behaviors can help. Is the bird diving often? When it flies, does it flap its wings slowly, quickly, or in between? When birds fly, keep your eyes peeled for color patterns on wings; the color of a duck's speculum (the bright patch on its flight feathers) is a good clue to identifying it.

Ring-billed Gull at nighttime roost with others in background

SEARCH FOR SEEDS

GREAT FOR:

- Sparrows, finches, buntings, chickadees, and more
- Finding birds in winter
- Irruptive birds such as siskins and crossbills

A MESS OF TALL, DRIED GRASSES might seem like an eyesore, but to many birds, it's a serious food party. The seeds of wild grasses, such as little bluestem, switchgrass, and prairie dropseed, are an important food source for migrating sparrows and buntings, especially in fall and winter. Look for movement at the base and border edge of grass clumps and clusters. Also check for small bird shapes among the tufts of grass; birds cling to the grass blades, often unintentionally swinging as they balance and stretch their bodies to reach the seeds.

In winter, the seed heads of dried flower stalks, especially those in the aster or sunflower family, are food magnets for birds like goldfinches, cardinals, and chickadees. Pine Siskins, nomadic finches, may suddenly descend on seed heads during a winter irruption (a sudden upsurge in numbers)—when food availability becomes scarce and birds are desperate to find it elsewhere. (This happened once in my Brooklyn patch; out of the blue, dozens of siskins were dangling from the seed heads of the flower field. They stayed for days!) Siskins also feed on seeds from pine cones, as do other nomadic, irruptive finches called crossbills, whose bills are specialized for extracting the seeds. In addition to finches, chickadees, and other boisterous seedeaters, nuthatches frequent pine cones—watch for their clingy movement in cone clusters.

Pine Siskin

23

CHECK AROUND THE EDGES

GREAT FOR:

Wrens, sparrows, thrushes, secretive ground-dwelling birds

AFTER A FEW DAYS on Hilton Head Island, I decided to walk a bit inland from the coastline, along an edge where the sand met some scrubby dunes and grasses. There, hopping on a fallen wooden fence, I found a trio of Savannah Sparrows, the only ones I had seen on my trip. They were taking advantage of easy pickings—seeds that had fallen from the dune grasses to the sand—and when I walked by, they ducked deeper into the grasses to escape my predatory feet.

"Edge habitat" is any place where the landscape changes. In a city, this could be a paved path through a wooded park; on a beach, a patch of prairie grass adjacent to the sand; in a forest, a clearing among the trees; on a lake, where the marsh meets the open water; and on an interstate, a ditch alongside the road. Some species of birds prefer edge habitat, as it offers them easier or more plentiful sources of food plus a quick escape to the safety of dense cover.

When in a forest or grassland, stop at any clearing to look and listen for birds in the bordering vegetation and below. In cities and suburbs, scan far ahead on empty paths and edges before you approach; use binoculars, if possible. In water habitats, watch edges of marshes or mangroves for birds leaving and entering or foraging just inside the border vegetation.

GOOD TO KNOW: *Many hawks take advantage of the edge, perching high alongside a road where they can easily detect a small mouse or vole scurrying out of median or shoulder vegetation.*

Ovenbirds often forage along the edges of park paths in New York City and strut back into denser leaf litter as people approach.

Curve-billed Thrasher

24

HANG OUT AT THE VISITOR CENTER OR PARKING LOT

WHILE SEARCHING FOR A STELLER'S JAY in Muir Woods National Monument in Northern California, Connor and I thought finding this reportedly common bird would be an easy task. But after a beautiful and tranquil two-hour walk through the woods, we still had not seen a Steller's Jay. A pair of Brown Creepers scaling some bark with their young fledglings fascinated us, as did a solo Common Raven perched at eye level and squawking while contorting its body. After spending quality time with a banana slug, we headed back to the park entrance and decided to check the gift shop. There we found the perfect present—a Steller's Jay hopping just outside the store's door.

This is not the first time we've found awesome birds at a park entrance. A Curve-billed Thrasher at Estero Llano Grande State Park in Texas, a Say's Phoebe at Madera Canyon in Arizona, and a Puerto Rican Tody in Cambalache State Forest in Puerto Rico all greeted us upon arrival. While many visitor centers have bird feeders and baths that attract various species, you'll also find a lot of birds hanging around those that don't have them. This might be due to the shelter or prime perching spots that surrounding buildings and signs provide, or possibly the fresh water that accumulates in parking lot puddles after rain. Whatever the reason, no matter how incredible the wooded trails or marsh in the distance appear, consider the visitor center and/or parking lot a great habitat for finding more birds.

GREAT FOR:

- Surprising discoveries of birds you really want to see (target birds) with minimal effort
- Local resident or specialty birds

Juvenile White-crowned Sparrow feeds on seeds.

25

VISIT GARDENS WITH NATIVE PLANTS

GREAT FOR:

- Maximizing the number of species you can find in a single location
- Resident and migrating songbirds, including sparrows, finches, thrushes, warblers, and vireos

EVER SEEN A MESSY, overgrown area that looks like someone is seriously slacking off on their gardening? You're in luck—birds like warblers, thrushes, catbirds, sparrows, and finches love this type of habitat, especially when it consists of native vegetation. Native scrubby bushes, weeds, and prairie grasses provide plenty of seeds, insects, berries, and sap, and places for birds to hide from predators. If you've been visiting a pristinely landscaped park with spacious lawns, peppered with non-native trees and ornamental plants, and are only finding House Sparrows, pigeons, and starlings, look elsewhere to find more bird biodiversity. Even an overgrown empty lot at the end of the block could be a better bet.

During spring and fall migration, many species of birds descend on native trees and shrubs that are crawling with protein-packed snacks like beetles and caterpillars. It doesn't matter if these plantings are in your backyard, your local park, or even the tiniest of green spaces down the block—they will attract birds. If a non-birdy park is your only option, getting to know the gardeners or staff and asking some questions about the trees and plants could start a nice dialogue. You might ask if they have ever considered planting milkweed, the host plant for monarch caterpillars and one of the most popular native "starter" plants.

Gray Catbird feeds on the fruit of native pokeberry.

Native oak trees, both large and small, are an especially big draw for migrating birds. In his book *Nature's Best Hope*, entomologist Douglas Tallamy points out that oaks support 897 caterpillar species, a huge source of food for both migrating birds and those raising young (most land birds feed their young caterpillars). Come fall, oak leaf litter (see tip #20) is chock-full of arthropods, an important food source for migrating and overwintering birds.

Native prairie grasses such as little bluestem and switchgrass attract sparrows, buntings, and finches that eat their seeds. On an elevated pathway in my patch, surrounded by tall switchgrass on either side, I found a rare Painted Bunting in the winter of 2019 that stayed for over two months; the endless supply of switchgrass seeds allowed the bird to snack all day long.

GOOD TO KNOW: *If you have a yard, it's easy to attract more birds with native plants—search for "native plant finder" online for lists of species that work well for your area.*

Yellow Warbler eats pink turtlehead.

Brown Creeper

26

BEWARE OF BARK THAT MOVES

GREAT FOR:

Nuthatches, creepers, woodpeckers, certain warblers

MANY TYPES OF BIRDS scale tree trunks, most notably woodpeckers, but also nuthatches, creepers, certain warblers, and even House Sparrows (which in my opinion can mimic the behavior of dozens of species, not only trunk crawlers!). Some catch the eye more easily than others that blend in seamlessly with tree bark. The Brown Creeper pictured here is an example of a bird that can be easy to miss—it's tiny (around 5 inches/ 13 cm from tip of bill to end of tail) and has brownish salt-and-peppery back plumage that looks like bark!

When you see trees with rough bark—full of insect-filled cracks and crevices—scan for "moving" tree trunks, especially during spring and fall when migrating birds are constantly foraging. Even if a bird is on bark right in front of you, its plumage might be too camouflaged to detect; move around the trunk to get a better angle on the edges and detect silhouettes. Also scan up the tree and along the limbs and branches.

Once you find a bird, you can get more quality time with it by knowing its preferred climbing technique. Brown Creepers generally scale up a tree; once they approach the top, they fly down to

Black-and-white Warbler

the bottom of an adjacent or nearby tree to continue the upward climb in search of food. Brown Creepers also tend to move from tree to tree in the same direction (I wait a few trees down the line to get photos of them!). Woodpeckers often scale upward, though sapsuckers (a type of woodpecker) often circle a trunk first, drilling horizontal holes that tap sap to trap insects for future snacking. Nuthatches can move in all directions over tree trunks and limbs but usually have their heads facing downward. As for House Sparrows, I've seen them do all the above!

GOOD TO KNOW: *Black-and-white Warblers of the eastern US are also notorious creepers; Yellow-throated Warblers, mostly found in the Southeast, also exhibit this foraging style.*

27 WATCH MUDFLATS AND OTHER MUDDY PATCHES

GREAT FOR:

- Shorebirds, including sandpipers, plovers, avocets, and stilts
- Large foraging flocks often numbering in the thousands (on large mudflats)
- Observing the intense foraging behaviors of shorebirds as they take advantage of low tide

HOW DOES MUD STEW SOUND for dinner? For shorebirds, it's a gourmet meal. When the tide flows out of bays and marshes, it exposes mudflats packed with crabs, snails, mollusks, and worms—even a nutritious goo of microorganisms called biofilm. Yummy! Next time you come across mud, especially near a coast or marsh, check for shorebirds digging their bills into the surface to pick out crunchy, chewy, and slimy snacks. Binoculars (see tip #7) are a big help, since the birds tend to blend in well with the muck, and some might be far off in the distance. (Many birders bring spotting scopes to large mudflats to view distant birds.) To maximize your finds at shoreline mudflats, check the tide times and head out at low tide, when more mud is exposed—or, even better, at rising tide, when birds are forced to congregate in a smaller area and will be easier to observe as they forage closer to shore.

Coastal zones aren't the only places to find birds slogging daintily through the mud; even small inlets and water edges of urban parks can attract birds, especially during fall shorebird migration (which, in the US, actually starts in summer and peaks in August). Always check your local muddy patches, including pond edges and puddles. A small tidal mudflat in my birding patch, usually frequented only by resident geese and ducks, recently turned up a rare Least Sandpiper.

GOOD TO KNOW: *In spring and early summer, you can also find birds gathering mud for their nests, especially Barn and Cliff Swallows. If you see swallows circling over a muddy area, watch as they land and fill their bills with sludge.*

Marbled Godwit on Hilton Head Island, SC

28

LOOK FOR BIRDS ON BALL FIELDS, EVEN WITH ARTIFICIAL TURF

GREAT FOR:

- Birds that hunt insects and small reptiles on short grassy fields and lawns, such as bluebirds, shrikes, and raptors like the Red-shouldered Hawk

- Finding Killdeer on the ground, and possibly their nest or young

- Discovering birds in a seemingly vacant landscape

GAME DELAY. Dapper nesting birds called Killdeer are responsible. Lacrosse season has been postponed and soccer matches moved so the birds can incubate and raise their young in peace. Named for the sound of its call, Killdeer frequent wide-open spaces like ball fields, beaches, and gravelly or cracked parking lots. They hang out in this habitat in winter (in the southern US) and nest there in spring and summer, choosing to lay eggs in a mere depression in the sand, gravel, dirt, grass—even artificial turf.

Always scan open spaces for Killdeer or other birds. At first glance, these habitats often appear devoid of wildlife, but you could be walking by dozens of birds without even realizing it. One winter on Hilton Head Island, I noticed an "empty" baseball field and did the obligatory bird-finding check. Sure enough, as I scanned the grass, I discovered over twenty-five Killdeer dotting the lawn. Staying put to observe them for a few minutes (see tip #4) resulted in even more birds—over a dozen Eastern Bluebirds flew in and landed to the right of the Killdeer. What had appeared to be a lifeless ball field in winter was full of birds anyone would want to see.

GOOD TO KNOW: *The "nests" of Killdeer are vulnerable to accidental trampling, but they use an injury-feigning display to try to protect their eggs and young. If a raccoon, snake, or human approaches, a parent attempts to lead them away by hobbling along with wings drooping, tail dragging, and reddish-brown rump feathers showing.*

Killdeer on baseball field

Song Sparrow

29

CHECK FOR OPEN PATCHES OF GROUND AFTER A SNOWFALL

GREAT FOR:

- Clear and close views of birds in places they might not normally forage
- Birds being less skittish as they stick close to and defend their limited food sources
- Observing birds in survival mode

AS MENTIONED IN TIP #3, cold weather can freeze fresh water sources that birds depend on; it can also blanket the dinner tables of many birds in several feet of snow. While birds that eat berries can dust off freshly fallen powder from trees and shrubs to get to the fruit, others that feed from the ground or below it face a dilemma for survival. American Robins and American Woodcocks both dig for earthworms, but it's the woodcocks that struggle more in snow—they can't survive on berries like robins, and they depend on earthworms and other invertebrates to survive.

When their food sources are inaccessible, ground-foraging birds often forgo a fruitless, energy-zapping digging expedition and instead seek small pockets of open ground. After a heavy snowfall, find more birds by scanning for dark openings where patches of dirt or grass stand out amid the sea of white, especially under trees and shrubs. Also check areas where ice is more likely to melt quickly, along cement paths, roads, and buildings.

Northern Flicker

Black-crowned Night-Heron removes invasive goldfish from park pond.

30 WATCH FISH PONDS LARGE AND SMALL

GREAT FOR:

- Herons and egrets
- An entertaining show as birds hunt fish (especially when they're having trouble getting down an oversize catch!)
- Catching a glimpse of swallows and swifts quickly dipping in for a drink (in summer)

THE LAST PLACE YOU MIGHT EXPECT to find North America's largest heron is in a tiny pond. But the 5-foot-tall (1.5 m) Great Blue Heron visits ponds so small it can barely take a single step before reaching the water's edge. Brightly colored goldfish swimming in shallow water below easily catch the attention of herons and egrets as they fly over. Many garden pond owners have taken steps to prevent these prehistoric-looking, sword-billed birds from stealing and snacking on their fish; deterrents include everything from piano strings crisscrossed over the water to motion-activated sprinkler blasts.

Ponds and streams in parks and nature areas, however, are often full of discarded pet goldfish and koi and their offspring. These have a negative effect on the local wildlife, stirring up sediment and disrupting native fish and amphibian populations. Luckily, herons and egrets help manage these invasive species, visiting ponds and picking out snacks.

Green Heron

These fish-fancying birds can be easy to miss at first glance; they often stand motionless, watching the water for the perfect catch. And when they do move, they're in full stealth mode, walking in slow motion so they can remain undetected by their prey. If you encounter a fish-filled pond or stream, scan the edges and shallows for a large wading Great Blue Heron or Great Egret; also look for their long necks peeking up through any surrounding water vegetation, especially tall marsh grasses and cattails. Green Herons are much smaller, better camouflaged, and fish from pond edges or low-lying branches and vegetation, so you'll want to scan low over the water's edge. Black-crowned Night-Herons do sometimes fish during the day, but more often they rest above the water; scan for their stocky and hunched vertical oval shapes or silhouettes in any branches 10 to 20 feet (3–6 m) over the pond.

GOOD TO KNOW: *Green Herons are adept hunters that sometimes place sticks or leaves in the water to bait or distract fish and then eat them.*

31

VISIT BERRY BUFFETS

GREAT FOR:

American Robins and other thrushes, waxwings, bluebirds, and catbirds, among others

WHEN YOU COME ACROSS a tree or shrub full of berries, you've found a favorite restaurant of dozens of birds, including robins, waxwings, catbirds, thrushes, and bluebirds. Birds eat all sorts of the small fruits, including sweet serviceberries, tart and tangy hawthorn, and fuzzy sumac. They even enjoy berries that are toxic to humans, including those of holly, pokeweed, and poison ivy.

Assess the berry situation in your location by regularly visiting fruiting plants to look for movement and listen for sounds. Birds can be found in these plants throughout the year, but you'll notice more in fall and winter as temperatures drop and insects and other invertebrates become scarce and tougher to find. If you come across birds feasting on fruit, watch and enjoy, then look for nearby plants offering the same food; once the current tree is plucked, they'll move on to the next. Knowing where the fruit is will also help you determine where large flocks of birds have landed, even if they're far away. When a group of over a hundred American Robins seemed to disappear into the trees of my birding patch, I headed to the dogwoods—which were fruiting their cornelian "cherries"—to find them.

GOOD TO KNOW: *Black-and-white Warblers, Common Yellowthroats, and other wood warblers are often seen wrestling with wriggly caterpillars, not plucking berries. Most must migrate south in winter to carry on their insectivorous lifestyles. But Yellow-rumped Warblers can tough out freezing winters, and some may stay; they're able to digest the wax that covers wax myrtle berries (aka bayberries) and get to the nutrients of the fruit.*

Cedar Waxwing feasts on a mulberry.

CREEP UP ON CATTAILS

GREAT FOR:

- Red-winged Blackbirds, often making food deliveries to their young nestled deep within cattails

- Chickadees and wrens

- Nesting ducks and their young (at the base of the cattails)

- Listening for rails that often remain hidden at the bottom of cattail marshes. Use the Merlin Bird ID app to help you identify birds like King Rails and Virginia Rails that call in cattails.

IF YOU SEE SOME CATTAILS, get ready to find some birds. Dozens of species, including blackbirds, chickadees, ducks, wrens, and rails can all be found hanging around them. In this marsh habitat, birds are often well hidden as they forage or nest low. To find them, use some of your other bird-finding techniques—stay a while, look for movement, and listen for sounds.

Some birds pick seed fluff from cattails to line their nest and make it extra soft and cozy for their young; hummingbirds, bluebirds, and bushtits fall into this category. Others weave nests low around the plant's sturdy stalks, the noisiest of these nesters being the Red-winged Blackbird, and the busiest being the Marsh Wren, known to build over a dozen decoy nests in a single breeding season.

Birds use cattail habitat in different ways depending on the season. During spring and summer, watch for birds flying or swimming in and out of cattail marshes—a sign they are building a nest or feeding their young. A bird pulling apart a cattail means they are just stopping by to pick up some fluff; track their flight path to find their nest. In winter, watch for birds snacking on these "hotdogs-on-a-stick." Chickadees sift through the fluff to feast on moth larvae, and American Goldfinches extract the seeds.

GOOD TO KNOW: *It wasn't until cattails were planted in my birding patch that Red-winged Blackbirds became frequent visitors. A year later a pair nested in the tiniest patch of the stalks, only about 6 feet (1.8 m) in diameter. Hooray for microhabitats!*

Black-capped Chickadee

Northern Emerald-Toucanet in Costa Rica

33

VISIT BIRD FEEDERS

GREAT FOR:

- Sparrows, finches, chickadees, nuthatches, cardinals, jays, woodpeckers, and more
- Observing entertaining battles and bickering as birds vie for the best feeder spots

WHILE THE SEEDS, fruits, sap, and insects found on native plants are important food sources for birds, you can definitely find birds at feeders, especially granivores whose diet consists primarily of seeds. Some parks and wildlife refuges may have bird feeders, often near their visitor centers. Before visiting a park, refuge, or nature center, try to find out if they have feeders and where, or ask staff when you arrive.

When we visited Colombia, our bird guide (see tip #74) planned a visit to a private *finca* with a hummingbird feeding station. While enjoying a tasty breakfast accompanied by coffee and hot chocolate, we observed dozens of colorful (and feisty) hummingbirds just a few feet in front of us. Restaurants in other birdy locations sometimes have feeders, so anyone can enjoy a meal while watching the birds have theirs. On a recent trip to Costa Rica, we visited several; one was packed with Fiery-throated Hummingbirds sparring over prime spots at the feeder. Another had a fruit feeder platform visited by toucans and brightly colored tanagers. But the most entertaining had to be one where two mammals of Costa Rica—a giant but cute rodent called an agouti and a type of weasel called a tayra—ran off with whole bananas from the bird feeder platforms.

Black-capped Chickadee (left) and Downy Woodpecker (right) share a feeder.

GOOD TO KNOW: *Too many starlings and grackles visiting your feeders? Try stocking them with safflower seeds. The oil doesn't taste that great to them—or to squirrels either. To prevent mycoplasmal conjunctivitis, an eye disease that affects finches (especially House Finches), thoroughly clean feeders every week or two.*

34

VISIT A BUSY URBAN PARK

GREAT FOR:

- Local resident songbirds

- Close-up views and higher numbers of migrating songbirds in spring and fall, foraging in smaller and more concentrated habitats

- Raptors hunting rodents and other small mammals

THE LAST PLACE you might expect to find more birds is in a packed city park. But if you look beyond the pigeons to the microhabitats that surround them—such as trees, shrubs, marshes, ponds, and lawns—you will find more birds. Best of all, you can get very close views of birds that, in larger wooded areas, can be difficult to observe, even with binoculars. (There's a reason "urban birding" is so popular!) Spring and fall migration are the best times to find the largest variety of species, as dozens stop at even the smallest of green spaces to eat and rest during their journey.

It's easiest to find these urban birds early, before a park gets too crowded, which is when they'll be foraging out in the open, on lawns and along walking paths. But with a little more effort, you also can find them throughout the day. As more people arrive, many migrating birds will stick around and continue the search for food but will move deeper into dense shrubs and higher up in trees. In New York City parks, for example, people continuously pass by low shrubs during migration, but many don't notice the activity right at their feet on the side of the path. An Ovenbird struts like a chicken and gobbles down arthropods while grackles and catbirds push aside leaves as they search for food. A bit higher up, at knee level, a Common Yellowthroat feasts on bright green caterpillars in a flowering shrub. Next time you visit a busy park, it's definitely worth a look and a listen; there are plenty of birds to be found.

Nashville Warbler in Brooklyn Bridge Park

Ruddy Turnstone

35

BUNDLE UP AND HANG OUT NEAR THE JETTIES

GREAT FOR:

Sandpipers, turnstones, surfbirds, oystercatchers, and dapper-looking ducks

WHILE DOING SOME WINTER BIRDING at Rockaway Beach in Queens, New York, I spotted a rock jetty off in the distance. I had heard that Purple Sandpipers—a bird I had always wanted to see—hung out on jetties in winter, so I made the trek. Like clockwork, there they were; they immediately impressed me with their agility and insouciance as the waves kept crashing in.

Filled with fish, seaweed, mollusks, and crustaceans, jetties are a giant sushi bar for birds. Shorebirds that frequent rocky jetties and coastlines in winter are called "rockjumpers." They move effortlessly around the rubble, dodging waves and staying afoot on slippery algae as they feast on mussels, limpets, crabs, and the like. Ruddy and Black Turnstones, Purple and Rock Sandpipers, Surfbirds, and Wandering Tattlers all fall into this category. Black and American Oystercatchers aren't quite as agile, but they are a sight to see, with their bills that resemble carrots and their colorful eye rings.

Sanderling

Decorative-looking ducks also visit jetties in winter, swimming right up to the bar for a blue mussel feast. These include Long-tailed Ducks, Common and King Eiders, and Harlequin Ducks.

GOOD TO KNOW: *Be safe! It's best to observe rockjumpers and other jetty visitors from the shore. Sea rocks are slippery, tides change, and "sneaker waves" can creep up on you. Enough said!*

36

VISIT A BEAUTIFIED DUMP

GREAT FOR:

- Grassland birds (even some species in decline, such as Bobolink and Grasshopper Sparrow)

- Uncommon birds needing open habitat

IN THE BIRDING COMMUNITY, dumps—most notably the Brownsville Landfill in Texas—are known as great places to look for birds, especially gulls, crows, and others attracted to food trash. But I'll spare you the scent, and instead recommend a more pleasant alternative: a capped landfill turned beautiful bird habitat. Landfills have a limited life span, and when they reach their end, cities and municipalities are often able to purchase the land at low cost and create parks and nature preserves. My favorite example is Croton Point Park in New York, where a landfill covered in grassland—a habitat that has been in serious decline for decades—attracts birds such as Bobolink, Grasshopper Sparrow, and Dickcissel. These now-beautiful garbage piles are found throughout the US, many in densely populated areas that were in desperate need of more community green space. Examples are Freshkills Park on Staten Island, New York; Washington Park Arboretum in Seattle; Red Rock Canyon Open Space in Colorado Springs; and the most aptly named Mount Trashmore in Virginia Beach.

GOOD TO KNOW: *Landfills are often the only undeveloped areas left in a city, so some governments start coordinating conversions while the landfill is still active.*

Dickcissel sings at Croton Point Park on the Hudson River.

37

WALK AROUND YOUR LOCAL COLLEGE CAMPUS

GREAT FOR:

- Local resident birds, such as jays, cardinals, and mockingbirds
- Owls, hawks, and other birds that hunt in open spaces

IF I HAD KNOWN THEN what I know now—wow! With their mature trees and open spaces, college campuses are great places for finding birds, but I had yet to become a regular bird-finder until I was well out of grad school. Apart from a Barred Owl I spotted at UNC–Chapel Hill (see tip #106; "Look at night"), I can't tell you one bird I noticed while attending college. Good thing it's never too late to go back to school—or at least visit the campus.

As I write, it's the start of winter—New Year's Eve, to be exact—and plenty of birds are being reported on eBird.org at universities across the world: Pacific Wrens at Portland Community College, Bushtits at the University of Colorado, Killdeer at Texas A&M, and a California Towhee at my alma mater, UCLA. While living in Florida, I frequented the wooded areas of the University of West Florida to find the impressive Pileated Woodpecker, and here in New York City, I enjoy walking through the Brooklyn College campus, where I've spotted Peregrine Falcons perched on building edges and Monk Parakeets nesting on light posts.

As you walk around a campus, look for nooks and crannies of good bird habitat, especially groups of shrubs with buds and berries and open areas behind buildings. Be sure to periodically scan the sky (see tip #8), as well as building corners, light posts, and fence tops overlooking open areas where hawks like to hunt.

GOOD TO KNOW: *Some colleges and universities carry out biological and ecological research in their very own campus wetland habitats, many of which are open to the public for wildlife viewing.*

House Finches are a common sight at colleges and universities across the US.

FINDING BIRDS

BIRDS

by the

CLUES THEY LEAVE

Tree Swallow nestling

38

CHECK FOR CAVITIES

GREAT FOR:

- Tree Swallows
- Bluebirds, chickadees, and nuthatches
- Woodpeckers and owls
- Cavity-nesting ducks

WE'VE ALL SEEN HOLES in trees and wondered if anyone was home—a bird, a bat, a raccoon. Often, we see nothing. But don't be discouraged; it's just that the entrances, exits, and peek-a-boos of wild animals are different from those of imaginary creatures we find in cuckoo clocks and cartoons.

Birds use cavities in living, dead, and partially dead trees as well as in buildings, nest boxes, and other structures. Some create holes for nesting and raising their young, while other "secondary" nesters lay eggs in holes created by others. They aren't lazy, they just don't have any carpentry chops. Try to picture a duck pecking out a hole, wood chips flying; it's just not going to work. Ducks such as Buffleheads, Wood Ducks, and Goldeneyes leave the head-banging to the woodpeckers—species with super-strong beaks and head-cushioning mechanisms—and instead go shopping for nest holes in spring. Other types of birds also take advantage of woodpeckers' craftwork. Tree Swallows often nest in cavities carved by sapsuckers who nested there first. Any time of year, birds may use holes to escape from predators or to rest or sleep.

So how do you find bird-filled cavities? Scan any dead trees or snags for holes and movement. Also be on the lookout for woodpecker cavities below shelf mushrooms that jut out from tree trunks; the birds take advantage of the easy excavation afforded by the decomposing wood. In spring and summer, a bright white drip-dry stain near the

Red-bellied Woodpecker

bottom of a tree hole means someone's probably home (this is fresh bird poop; see tip #39). Baby birds are likely inside waiting for a parent to drop in with a juicy caterpillar or iridescent dragonfly. You might hear them faintly whining, or they may be quiet until the parent is near, at which point things get loud. Walk past the hole and watch from afar for the parent to arrive. Another clue to an active hole is fresh wood chips or shavings around or below the opening—a sign that some bird has been hard at work. If you live near the hole, check it regularly for the grand entrances and exits of a woodpecker, or the more subtle comings and goings of a tiny nuthatch. And make sure your eyes are not deceiving you; some birds, especially screech-owls, are so well camouflaged it can be tough to notice them perched right in a hole opening.

GOOD TO KNOW: *Through their carpentry, woodpeckers create habitat (cavities) and expose food resources (sap, for example) used by many other birds, mammals, and insects. Because of this important role in sustaining their surrounding ecosystems, they are considered a "keystone species" of woodland habitats.*

Below the roost of a Black-crowned Night-Heron

39

CHECK ABOVE AND AROUND BIRD POOP

GREAT FOR:

- Herons, egrets, cormorants, owls
- Finding nighttime roosts of resident birds such as starlings or grackles

I NEVER THOUGHT I'd be spending time trying to get the perfect photograph of bird poop, let alone recommend you go looking for it. But what can I say? Birding changes you. So here we are with another great way to find more birds—keep your eyes out for bird poop, especially the "whitewash" type that stains branches, leaves, rocks, and ground. A large—say, 6 to 12 inches (15–30 cm) in diameter—dried pool of the stuff on the ground below trees can often mean a heron or egret is roosting there. Scan the branches directly above the excrement; even if there is no bird, one may return during its normal roosting time. (This is a great way to find Yellow and Black-crowned Night-Herons, since they roost during the day.) Check branches of pines and other dense trees for whitewash left by owls. (While you're at it, scan for owl pellets under the tree—small, compressed ovals of fur and tiny bones, which owls regurgitate after digesting nutrients from their small-mammal meal.)

A lot of scattered whitewash under a tree or shrub likely indicates a nighttime roosting site for a large group of birds, especially starlings or grackles. You'll probably hear them being very chatty when you approach them in the evening—a good warning to walk around rather than under the tree.

Cormorants are notorious for making a mess, or, depending on how you look at it, art. A blue tennis court provided the perfect backdrop for the excrement of a resident Double-crested Cormorant. I was worried I had missed my chance to photograph it after a heavy rain, but it was still there, shellacked to the surface.

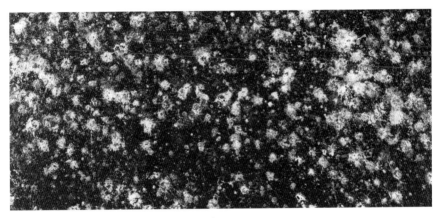

Artwork of Double-crested Cormorant on tennis court canvas

GOOD TO KNOW: *It's a common belief that getting pooped on by a bird is a sign of good luck. This helped me laugh off an embarrassing pigeon incident as I exited a local coffee shop.*

Baltimore Oriole drinks nectar from Paulownia bloom.

40

SMELL THE SCENT OF FALLEN FLOWERS, AND CHECK THEM FOR SLITS!

GREATE FOR:

- Orioles
- Hummingbirds
- Flowerpiercers (found in South and Central America and Mexico)

THE IMAGE OF A HUMMINGBIRD enjoying nectar from a flower is one we all know and love. The fact that these feathered gems provide important pollination services while imbibing makes the scene even more special. But then there are the nectar robbers: birds that commit "floral larceny"—a term I am not making up—by piercing or biting the bases of flowers to sip the nectar. These include certain species of hummingbirds, such as the Reddish Hermit of South America, but birds aren't the only ones committing this crime; some bees, bats, and other mammals are also guilty.

In spring and summer, be on the lookout for clusters of fallen flower blooms. After picking one up to enjoy its scent, inspect the base for signs of forced entry like a slit or a small hole. A slit often means an oriole or flowerpiercer has robbed the nectar; check plants above for still-attached flowers and birds foraging among them. Each summer

Evidence of nectar-robbing

in my local birding patch, Brooklyn Bridge Park, the ground at the top of Granite Prospect winds up covered in the sweet-smelling purple flutes of the Paulownia tree for a week or two, when it's easy to find slits in fallen blooms and several Baltimore Orioles imbibing in the trees above. (Smaller holes might be the work of bees, but it's still worth a look up.)

GOOD TO KNOW: *Charles Darwin believed that nectar-robbing had a negative effect on plants and prevented pollination, but that theory has since been disproved. Removing nectar through a flower's base can still result in pollination. In their paper on queen bumblebees exhibiting this behavior, a team of researchers at Hokkaido University suggested it be called "robber-like pollination."*

House Sparrows bathe in the dust.

41

LOOK OUT FOR BIRD SPAS . . . DUST BATHS

GREAT FOR:

- House Sparrows
- Roadrunners
- Chickens
- Witnessing the curious behavior of bathing in the dirt

I CAME ACROSS a letter to the editor from a woman asking about mysterious "mini crop circles" that had appeared in her yard. She was baffled as to what creature or otherworldly phenomenon was responsible for the curious dirt formations. It turns out that when she wasn't looking, birds were bathing—in the dust.

Over two hundred species of birds have been known to take dust baths; most do so in arid climates where a more traditional bath is not an option. But House Sparrows, the most notable dust bathers in North America, engage in the practice all the time. I find them nestled in depressions, fluttering their wings wildly, creating dust clouds that cover their feathers and settle down to their skin, absorbing excess oil and displacing mites. The practice keeps their feathers in top-flight condition.

Other birds that get clean by getting dirty are wrens, larks, thrashers, even roadrunners, quail, and chickens. (People who keep chickens often provide them with an area to dust bathe; some even pamper them with herbal dust-bath mixtures.)

If you come across shallow depressions in loose dirt or sand, you may have found a bird spa! Since birds engaged in this cleansing ritual can be quite skittish (a bathing bird is more vulnerable to predators), observe from a hidden location and wait for birds to arrive. Employ tip #2 ("Look for movement"), and be sure to scan any areas of open loose dirt or sand, including empty sand volleyball courts or ball-field dirt.

The spa is empty.

42

FOLLOW TRACKS

GREAT FOR:

Herons, shorebirds, ducks, and pigeons

I NEVER THOUGHT of following bird tracks to find birds until I unexpectedly came face-to-face with a Great Blue Heron after tracing its giant anisodactyl tracks (more on this below) in the sand. There I was, playfully following in the footsteps of some large bird, when the tracks suddenly ended in a dune surrounded by vegetation. As I raised my gaze, my eyes met with those of the 4-foot-tall (1.2 m) living dinosaur standing right in front of me.

Hunters, our allies in the conservation of bird habitats (and one of the main reasons some wetland bird numbers are increasing) use bird tracks all the time to locate birds; this is one of the reasons they like to hunt in the snow. Why not add this technique to our ever-growing bird-finding tool kit? The next time you see bird tracks, view them as a clue to the presence of birds and see where they lead you.

A track left by a Snowy Egret while foraging along the shore

GOOD TO KNOW: *Bird tracks exhibit various toe arrangements. The anisodactyl tracks of many perching birds, such as sparrows, thrushes, doves, and herons, show the most common toe arrangement in the bird world, with the first toe (the hallux) facing back and the other three facing forward. Birds like woodpeckers, owls, and roadrunners sport the zygodactyl arrangement of two toes in front and two in back.*

Great Blue Heron

Yellow-bellied Sapsucker maintains its phloem holes.

43

KNOW YOUR WOODPECKER DRILL HOLES

GREAT FOR:

- Sapsuckers busily drilling and maintaining their sap wells
- Birds that feed on sap, such as kinglets, warblers, and hummingbirds

WOODPECKERS ARE SOME OF THE MOST skilled carpenters of the bird world, but how do you catch one in the middle of its impressive work? Lucky for us these birds leave telltale signs of their presence in the form of drill holes and cavities. The sapsuckers are the easiest to trace: Yellow-bellied Sapsuckers of the eastern US and Red-breasted, Red-naped, and Williamson's Sapsuckers of the Pacific Northwest. If you encounter neat rows of deep, round holes on a tree trunk, you've found the work of a sapsucker. These are called "xylem holes" because they reach the xylem layer of the tree bark, tapping into the sweetest sap of the tree. Sapsuckers also artfully carve out "phloem holes" in rectangular-patterned rows, which, as you may have guessed, reach the phloem layer of bark. These are not as deep as the xylem holes and to keep them flowing requires constant maintenance. I watched the Yellow-bellied Sapsucker pictured on the previous page for well over an hour as it tended to its phloem holes.

Xylem holes

Through all their hard work, sapsuckers create their very own bar with a variety of flowing taps, some sweet (xylem) and others bitter (phloem). They visit their "sap wells" regularly and lick them with the bristly tips of their specialized tongues. But how do they get their protein? Easy—they pick off insects that get stuck on the sap. Other bird species—hummingbirds, kinglets, and warblers, to name a few—also visit in search of an energizing treat, but the sapsuckers are quick to defend their territory and drive them away.

Visit trees that have been "sapsuckered" regularly to find these woodpeckers hard at work, along with other birds and insects visiting the taps.

GOOD TO KNOW: *Wood-boring insects also make holes in trees, but these holes are not patterned in rows and will appear more random.*

44

LOOK FOR FALLEN FEATHERS, A SIGN OF NEARBY PREDATORS

GREAT FOR:

- Raptors and their prey
- Discovering raptors' nests and favorite "plucking perches"

FINDING A FEATHER is common enough, but finding a scattered group of them can lead you to a raptor, many of which have plenty of smaller birds in their diet. Different birds of prey have different "food prep" techniques, but most pluck at least some of the feathers before and during the meal. If you come across a mess of feathers, there's likely a raptor in the neighborhood. Note the perch above the plumes and check it regularly for hawks or falcons; scan the skies (see tip #8) for soaring birds of prey.

If you're lucky, you might come across a banquet in progress, with feathers falling from above; look up to find a hawk or falcon feasting on its prey from its "plucking perch," such as a tree, post, or building. Some hawks also consume their food on the ground, and you might notice feathers floating in the wind. After the plucking process, if you see the bird fly off with food hanging from its bill, it could be delivering it to its young. Track the bird in flight (see tip #102) as far as you can for a chance to find a raptor nest.

American Kestrel plucks feathers from its White-throated Sparrow prey.

FINDING
BIRDS

through

TECH
SUPPORT

✳
✳✳

45

LET MERLIN LISTEN

GREAT FOR:

- Finding backyard birds you had no idea were there!

- Discovering what birds are present in a location so you know what to look for

- Learning birds' songs and calls

- Finding birds when you have trouble hearing

WALKING THROUGH THE CONSERVATORY Garden in New York City's Central Park, I followed the distant sound of what I thought was a Common Yellowthroat warbler. But as I approached, I realized that while the notes had a similar cadence to the yellowthroat, they were raspier and a bit electronic sounding. The bird was hidden in a manicured hedge, and I had no idea what it was. But Merlin did. I fired up the app's Sound ID and it immediately reported an uncommon Mourning Warbler (a bird in the same genus as the yellowthroat!). I was determined to get a look and found the bird singing low in a tree on the other side of the hedge.

Merlin is a free app for iOS and Android that identifies bird sounds and photos, and even includes a field guide tailored to where you are. It's awesome for finding more birds—start Sound ID and find out what Merlin is hearing in your backyard or local park!

GOOD TO KNOW: *Merlin is highly accurate, but so are mockingbirds. If all the birds Merlin reports seem to be coming from the same source, check for a Northern Mockingbird on a nearby light post or fence. You can also let Merlin listen to your neighborhood mockingbird to find out its repertoire and learn bird songs in the process!*

Mourning Warbler in Central Park

46 SCAN EBIRD BAR CHARTS

GREAT FOR:

- Knowing what species to expect in your area now, next, month, next summer, etc.

- Quickly scanning which birds are common and which are rare in your region

- Bird-finding on trips and vacations—you'll know what birds to watch for during your stay and where to find them

EBIRD.ORG IS ONE OF THE LARGEST crowdsourced biodiversity databases in the world. As citizen scientists, you and I and people in all corners of the globe can enter our bird observations into eBird. Our data is then publicly available for use by scientists and researchers working toward the conservation and protection of birds and their habitats. I love everything about eBird, but if I had to choose one feature I could not live without, it would have to be Bar Charts. Don't let the name fool you—they are super exciting and eye-opening. If you like birds, it will change your life. So how do you use them? Visit eBird.org/explore, look for "Explore Regions," and enter your county in the search box. This will take you to the "region" page for your county. Look for "Bar Charts" and give it a click. Here, you'll see a scannable list of monthly distribution charts for every species ever reported in the county. Scroll down, staying focused on the column for the current month, and you can easily tell which bird species are in your county right now. That will give you plenty to look over for your county, but there are Bar Charts (and region pages) for areas other than counties, including states, countries, and—most important—hotspots such as your local park!

GOOD TO KNOW: *Get ultra-local and visit the Bar Charts for the park nearest you. If they look a bit sparse, and there aren't a lot of observations, guess who can change that? You! Happy eBirding.*

Yellow-rumped Warblers migrate north earlier in spring and head south later in fall than most warblers. You can see this at a glance with eBird Bar Charts.

Scarlet Tanager

47

VISIT BIRDCAST.INFO DURING MIGRATION

SPRING AND FALL MIGRATION are the best times to find more birds, as hundreds of millions migrate through North America and make pit stops to seek food, shelter, and rest. Most songbirds migrate at night, taking flight thirty to fifty minutes after sunset. That's when the risk of being attacked by their aerial predators decreases; the temperature cools, so they require less caloric energy to power flight, and the setting sun and stars help them to navigate toward their destination.

If you live in the forty-eight contiguous United States, you can get a three-day forecast of the intensity of bird migration in your area thanks to weather radar stations with publicly available data and the analysis of that data by the team at BirdCast (birdcast.info), a collaborative project from the Cornell Lab of Ornithology and its partners (and one of the most exciting I've had the pleasure of working on at the Lab). It's long been known that birds were appearing on weather radar—showing as groups moving faster than the wind, or against or through it—but only recently did the intense computer processing power required to analyze this data become available.

As if a three-day bird migration forecast weren't exciting enough, birdcast.info also lets you view real-time maps and charts showing the direction, intensity, and number (often in the millions) of migrating birds as they move through the night. BirdCast's tools tip you off to the days when you're most likely to find more birds in your park or backyard, and when you will reap the greatest rewards from employing the many techniques in your bird-finding tool kit.

GREAT FOR:

- A chance to witness "bird fallout" after a storm
- Knowing when you can find the highest numbers of migrating birds in your area
- Experiencing the magic of migration!

GOOD TO KNOW: *Spring and fall bird migration don't line up exactly with the true seasons. BirdCast's "spring" reporting starts in March and lasts into June; "fall" migration kicks off in August and continues into November. That said, the birdiest times are right in line: In spring you'll find the most birds in mid-May, and in fall, during September and October.*

White-throated Sparrow

48   LEARN THE SOUND OF A BIRD NEAR YOU

WHEN I PLAY THE MELODIC SONG of my favorite bird—the White-throated Sparrow—for students or friends, many immediately recognize the sound, though they never knew where (or who) it was coming from. (It's not surprising, since the bird is well camouflaged in the leaf litter where it forages.) It can be tough to identify songs and calls of birds, but knowing what to listen for makes it easier.

Start by learning the sound of a common bird in your neighborhood. Use eBird's hotspot explorer (eBird.org/hotspots) to find the one nearest you; you'll land on a list of birds seen there recently. Click on a few of the birds and listen to their songs, then choose your favorite. Listen to it again several times as you look at the photos of the bird. The next morning, and every morning, listen for the sound of the bird. When you hear it, visualize the bird singing; if possible, watch it as it sings. Repeat this technique with other birds to build up your knowledge of local bird sounds. The more songs you learn, the easier it will be to learn another. (Find out what birds are common throughout the year by clicking on the Bar Charts link on the eBird hotspot page.)

Red-winged Blackbird

GREAT FOR:

- Getting on the fast track to easily identifying the sounds of resident birds near you

- Fine-tuning your ability to detect more birds in the area, even those vocalizing at a distance

- Finding masters of camouflage that you didn't know were there

GOOD TO KNOW: *The Merlin app's Sound ID is highly accurate. Use it to your advantage in improving your birding ear. First, try to identify all the birds you're hearing, then use Merlin to verify any you're not sure about. Review any species that Merlin heard and you didn't, and add those to your list to learn and detect next time.*

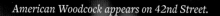
American Woodcock appears on 42nd Street.

49 TUNE IN TO SOCIAL MEDIA

WITH MILLIONS OF PEOPLE finding and observing birds in the US alone, it's easy to tap into what's happening in your local birding scene through social media platforms. Many people, including me, post photos of birds taken in their local parks and cities. Following these types of accounts (or joining such groups on Facebook) will clue you in to bird species in your area and often give specific information on where to find them. Many will also tip you off to rare bird sightings.

To find the most helpful accounts to peruse and follow on your preferred social media platforms, use the search feature and play around with different search terms. Enter your city, county, or local park name, along with terms like "birding" or "rare birds." I searched "birding Marin" on Twitter to see what I might find in my old home county; an extremely rare Ruff was reported at Inverness Park, and the post included a link to the eBird checklist with seventy-seven other species! On Instagram, I entered "#portlandbirding," which turned up a rare Northern Waterthrush at Whitaker Ponds Natural Area and a gorgeous Lazuli Bunting at Powell Butte Nature Park. (Hashtags are popular on Instagram, so you'll want to search for things like "#marinbirds" or "#brooklynbirds." Placing the city or county name first usually returns better results.)

I recently followed a Twitter lead for a chance to spend quality time with a quirky bird—an American Woodcock—that was being seen at Bryant Park in Manhattan. When I ascended from the subway, it was impossible to miss the large crowd watching the best show in town: this woodcock strutting to and fro on 42nd Street.

GREAT FOR:

- Rare birds that are outside of their expected range
- Local "celebrity" birds, like woodcocks doing a funky dance or hawks hunting downtown pigeons
- Sharing bird sightings with others in your community (so they can find more birds!)

GOOD TO KNOW: *Sharing bird sightings on social media is a great way to contribute to and network with the birding community. Keep in mind that it's best to refrain from posting exact locations of "sensitive species" such as owls or nesting birds, especially in big cities where dozens of birders might descend on the location and cause stress to the animals. (Some groups and channels have rules that prevent such postings.)*

Laughing Gull and lined seahorse

50 TAKE PHOTOS

IF YOU'RE TAKING PHOTOS of a bird, you've already found it, right? Not necessarily. I spent a while snapping away at a Bald Eagle, but it wasn't until I reviewed the photos later that I realized there were other birds in the scene. Way off in the bottom corner of one photo was a Peregrine Falcon carrying bird prey in its talons, being pursued from above by the eagle. (Bald Eagles are known for pirating prey.)

To find more birds (and other curious creatures), take photographs of flocks of birds in flight, in water, or on the ground. Review and zoom the photos and you might find a bird you didn't even know was there. Often, something is perched so far off it's tough to tell if it's a bird, so use your phone or camera zoom to help. I do this all the time; sometimes my zoom finds me a bird—usually a raptor—perched on a distant bridge or spire. Other times, I get a fake plastic owl or a jutting fixture, and after a laugh, I hit delete.

My most recent photo-taking bird-finding success was in Brooklyn Bridge Park. In the distance, flying north above the East River, what I thought was a trio of Canada Geese caught my eye. But something about them looked different. I struggled to capture a photo from so far away but was glad I did. When I zoomed in on the horribly blurry photo, I could see that they were Great Cormorants, birds I'd never seen in over ten years of birding there.

One of my favorite discoveries by photo was not of another bird. Noticing that a Laughing Gull was after some creature in the water, I hoped to capture a photo of it with a crab dangling from its bill. But after reviewing my photos, I realized I had captured something even more amazing—the gull and a lined seahorse staring at each other in the water.

GREAT FOR:

- Finding additional birds and other wildlife you didn't know were there (like the seahorse!)
- Detecting something different in seemingly same-species flocks
- Birds you can't identify. After you snap, upload the photo to the Merlin Bird ID app. (You can also check your field guide or ask a friend for help later.)

Trio of Great Cormorants found by taking a bad photo.

Yay! Vermilion flycatchers often perch low and out in the open.

51

VIEW PHOTOS FOR HABITAT SCENES AND CLUES AT MACAULAY LIBRARY

GREAT FOR:

- Visualizing how the bird will appear at a specific location
- Getting detailed habitat clues for specific birds you want to find
- Well-hidden or camouflaged birds

IT'S EASY TO LEARN what birds are in your local park by using eBird.org. (See how in tip #52.) But what if you're at an eBird hotspot and still can't find any of the birds you came to see? Look for clues in photos from that very location! In addition to submitting checklists of birds they've found, many "eBirders" upload accompanying photos that are extremely useful in letting you know what to look for when you visit. These photos become part of the digital collection of the Cornell Lab of Ornithology's Macaulay Library—which also contains photos, audio, and video of amphibians, fish, and mammals. Visit search.macaulaylibrary.org and enter the name of the hotspot for where you are. Sort the photos by newest first to see all the recent photos people have added to their eBird checklists. Enter a species name to home in on habitat clues for any of the birds you want to see. (Don't be concerned with photo quality; some of the worst photos provide the best habitat clues!) Scan the results, absorbing the scenes surrounding the bird. Is it perched out in the open or peeking out from the edge of a path or marsh? Is it feeding on berries from a certain type of tree or shrub? Diving? You get the idea.

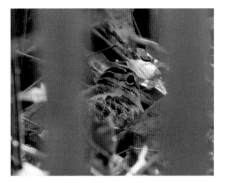

Oh no! American Woodcocks are masters of camouflage.

It's even better if you have a chance to look at photos before you head to a hotspot. (Before each of my trips, I search for photos of birds I'm really hoping to see, such as "life birds" or rarities.) Think of it as a bird texting you a photo (or message) of exactly where they're sitting in a restaurant so you can find them when you get there. Some like to lunch out in the open, others in a densely hidden picnic spot. When you arrive, keep an eye out for habitat scenes and landmarks (for example, signs, flagpoles, buildings) like those depicted in the photos. Many hotspots span a wide area, so it's a good idea to read photo comments for specific directions or location information; also look for people with binoculars gathered around the expected habitat.

GOOD TO KNOW: *Contributing your bird photos and audio to Macaulay Library will not only help others find more birds, it will aid in research and conservation. It's easy—just upload them to your eBird checklists and they will become part of the collection.*

Want to see a Greater Roadrunner?
eBird Explore can show you where to go.

52

USE EBIRD.ORG/EXPLORE

THE "EXPLORE" TOOLS AT EBIRD.ORG will help you find any bird you can think of; they will also help you discover fascinating species you probably wouldn't have conjured up in your wildest dreams. As of this writing, you can search for species and regions at the top of the screen at eBird.org/explore. Be sure to scroll down further so you don't miss out on a host of other useful tools, including "Hotspot Explorer" to see the birdiest places near you or anywhere; "Bar Charts" to know when and where to find birds (so ultra-useful in bird-finding that I gave them their own tip; see tip #46); "Species Maps" that show points for all locations where a species has been reported; and "Targets," which generates a list of birds you have not yet seen in an area.

Start by typing in your state or county in the "Explore Regions" search box and select the closest match. The page for the location/region loads. Recent bird sightings are listed, and you'll quickly notice dozens of other bird-finding resources just on this page. The best part is that they are all *specific to the location*. You can even print a checklist of species found there to take with you when you visit. (This feature was the very first thing I worked on when I started at eBird in 2016!)

From the region page, click "Hotspots" and drill down to one near you. You'll be able to find out what birds to look for when you visit your local park this week and what birds might not show up until winter. Keep in mind that many of the birds found in nearby hotspots can also be found right in your neighborhood, especially if it contains similar habitats. (My Brooklyn block doesn't have native prairie grasses, so I don't see Song Sparrows; it does have exposed perches, so I do see hawks.)

I use the explore and region pages all the time, whether it's to check on recent sightings in my patch, to make a list of birds I want to find on my next trip, or to see where in the world I can find a super-cool bird I just saw a photo of or read about. Make eBird.org/explore a go-to resource for your bird-finding needs.

GREAT FOR:

- Planning trips to find birds you really want to see or haven't yet heard of

- Detailed location information on where to find specific birds

- Knowing what birds have been sighted in your area and the best times of year to find them

- Signing up for rare bird alerts

FINDING
BIRDS

while

DOING
SOMETHING
ELSE

American Crow

53

GO TO YOUR LOCAL SUPERSTORE

GREAT FOR:

- Large flocks of blackbirds and starlings
- Finding birds while waiting for a friend to do their shopping
- Finding House Sparrow and other birds' nests in funny places—behind store signs or spikes meant to keep birds away

AS WE PULLED UP to Walmart to get some rotisserie chicken (yes, I eat chicken), I noticed some nice habitat surrounding the parking lot. I stayed and birded the superstore surroundings while Connor went inside. I was surprised to find Fish Crows dominating the lot, calling *uh-uh* from the trees and light posts. I only saw a single grackle—Boat-tailed—which, based on previous parking lot birding in Texas, I had expected to be more prevalent. Circling both high and low were dozens of Turkey Vultures and a few Black Vultures. An Osprey sat on its nest atop a light post while its mate called constantly as it circled over the lot. As I walked around the perimeter, I came across many White Ibis—thirty-four, to be exact—foraging in a ditch under some good tree cover. A trio of Eurasian Collared-Doves flew overhead and perched on cables at the end of the parking lot. Off in the distance, I heard a Blue Jay calling and a Northern Mockingbird running through its repertoire. Just then I saw my favorite bird—Connor—exiting the store. It was time to go, but it had been a nonstop, action-packed birding adventure in the parking lot of the Cape Coral Walmart.

Superstore habitats can vary widely, of course. Some may be adjacent to a gorgeous wetland, while in others you will be hard-pressed to find the saddest of shrubs. But there will be birds; superstore parking lots are buffets for birds that take advantage of discarded food trash. Check the tops of all light posts in the lot, look for movement around the perimeter and near the garden center, and scan the sky. In Texas, be sure to catch the sunset grackle shows at grocery, superstore, and mall parking lots, especially those with large trees where the birds roost at night. Some lots host thousands of these blackbirds (which are understandably considered a plague rather than a spectacle by some). If they're around, they won't be hard to find; you'll hear their wild mechanical-sounding weirdness all around.

An Osprey perches low in a superstore parking lot.

BIRD FROM THE BACKSEAT

GREAT FOR:

- Hawks and falcons perched on poles
- Flycatchers and shrikes perched on power lines
- Soaring vultures
- Sandhill Cranes foraging along roadsides (especially in Florida)

WHAT DO THE FOLLOWING birds have in common: a Red-tailed Hawk in Oregon, a Harris's Hawk in Texas, an Osprey in Florida, and a Merlin in New Jersey? They're birds you can see from the backseat of a car, perching on light posts and utility poles. These raptors rule the roadways from above, scanning for small mammal prey in the form of rodents, mice, voles, and rabbits. The lower vegetation level along roads and medians makes it easier for them to spot a snack, and the fresh roadkill options aren't bad either. Red-tailed Hawks are the most common highway hunters; on some road trips, I have seen one perched every few miles along major highways and interstates. In Texas, it was the Harris's Hawks that were perched on posts along many local roads. There's even a bird named a Roadside Hawk, found in parts of Mexico and Central and South America; I enjoyed spotting it on posts and cables in Colombia as my bird guide (see tip #74) took care of the driving.

Birds in flight can also be easy to spot from cars. The only bird I have ever seen in the state of New Hampshire is a Bald Eagle I spotted from the passenger seat as we passed through the state for a mere minute. Soaring vultures are such a common sight on many road trips that they have their own tip (see tip #62).

Birding from the backseat also offers non-raptor surprises that vary from region to region. For example, in Texas, Scissor-tailed Flycatchers are easy to find perched on power lines from spring to fall, and in Florida, Sandhill Cranes can be found year-round foraging along roadsides.

Sandhill Crane

WORK OUTSIDE

GREAT FOR:

- Wrens, mockingbirds, jays
- Neighborhood raptors
- Backyard birds

I'M HERE IN GEORGIA over the Christmas holiday, staying at Connor's family's beautiful place with a lake out back. On their deck, I notice a raised table and chairs overlooking the lake; this bird (me) has found its habitat. I don't see any birds, but I set up to write outside. Though it's cold, the sunlight warms me through the brisk winter air.

As I get to work, things start to happen; I see something large arriving out of the corner of my eye—it's a Great Blue Heron that lands in the shallows of the lake. A Carolina Wren starts to sing, and crows start to caw. I'm finding birds without even trying. As I raise my head to the sun, a raptor flies in and lands in an oak on the lake's edge. I'm not used to the birds down here, but after observing it for a while and noting its features, I realize it's a Red-shouldered Hawk looking down at me and my laptop. An hour passes; I hear chips and chirps here and there, including an Eastern Phoebe and a Northern Cardinal. Once again, a large wader catches my eye; this time it's a Great Egret. There is nothing like working outside, even in the dark. I not only hear the hoot of an owl, I also hear the gnawing vocalizations of the local beavers.

Carolina Wren in a Georgia backyard

56

VISIT FARAWAY FRIENDS AND FAMILY

GREAT FOR:

- Life birds (new species for your life bird list!)
- Observing common resident birds of another area
- Backyard birding

WITH OVER TEN THOUSAND bird species on the planet, no matter where you go, you'll have a chance to find more birds. Visiting faraway friends offers a bonus of encountering birds you have never seen before ("life" birds).

While visiting my friend Christine in Paris, I watched each morning from her beautiful apartment as Common Swifts wove through the quaint angled streets of the 3rd arrondissement. I took a pre-breakfast stroll to a tiny park down the block, where I saw a pair of Common Wood-Pigeons building a nest, a Eurasian Blue Tit, and a Eurasian Blackbird. Visiting my friend and fellow bird-lover Peter in Ecuador made for a bird-finding extravaganza. With his help, I met the Masked Trogon, Golden-breasted Puffleg, Flame-rumped Tanager, and over *one hundred* bird species!

Before traveling, visit eBird.org/hotspots and enter your friend's or family member's address in the "Location" field. There, you'll be able to see nearby birding hotspots and drill down to see a list of the birds you can find at the time of year you'll be visiting. Happy travels!

Masked Trogon

American Robin

57

GO FOR A WALK IN THE RAIN

- Non-timid earthworm eaters like American Robins and Rusty Blackbirds

- A chance to see many birds fly out from their perches to feed when the rain stops

WHEN I LEAD BIRD walks, they are always rain or shine. I have found this option much easier than coordinating a walk cancellation, and there are plenty of birds to see in the rain, often exhibiting different and more interesting behaviors than on dry days. I've seen Song Sparrows mating, an American Robin fledgling sticking its little yellow tongue out to drink falling raindrops, and a fluffy trio of Barn Swallow fledglings huddling together in an apparent effort to ward off a slight chill (pictured on page 5).

While wind causes many birds to take cover in dense and low vegetation—and makes them difficult to find—rain makes finding certain birds easier. Some

Rusty Blackbird

perch in trees and stay put rather than flying off the minute you lay eyes on them; others become more active around a food source that has been concentrated by the rain—like earthworms. Species that like to feast on earthworms are out and about, hopping and walking through soggy ground on lawns or in gardens, taking advantage of the many worms that have made their way to the surface. I once watched a Rusty Blackbird devour worm after worm at an alarming rate—much faster than the many American Robins that surrounded it. Every minute or so, one of the robins would notice this blackbird's incredible bounty and chase it away through the muck; unfazed, it would just continue its constant gobbling from a few feet away.

Another benefit of rainy-day birding is that the birds taking cover often "wake up" when there's a break in the rain. At first, it will be tranquil when the rain stops, but then you'll start to hear many birds singing and see them take flight.

58

TAKE YOUR KIDS TO THE PLAYGROUND

GREAT FOR:

- Getting to know the resident birds of the playground, like jays, cardinals, mockingbirds, and more

- Hawks and falcons, especially in playgrounds surrounded by large lawns or open areas where they like to hunt

- Migrating songbirds in spring and fall

- Kids learning about birds and nature firsthand

JUST A COUPLE OF BLOCKS up the street there's a cute little park with the scariest red-eyed yellow spider—made of metal—for kids to climb. While it takes only a minute or two to complete a lap around the park's circular brick walking path, the shrubby hibiscus patch on the inside and the giant elm and linden trees on the outside make for hospitable bird habitat. Sapsuckers, nuthatches, warblers, and hawks all appear on my "park list" for this playground that I visit only occasionally.

But I never knew playground birding was a thing until I heard the term while attending a Zoom discussion where mom birders shared ideas for how to fit birdwatching into their busy schedules. Turns out there are plenty of parents looking for birds on their daily visits to the local playground. And the best part is that their kids are involved in the effort, keeping a bird list and upping the playtime fun factor.

Scope out your local playground's bird-finding opportunities. Those with more trees and vegetation might attract more birds, but one in an open area could be a great place to spot the local hawk perched on a post and hunting over a lawn or field. Keep a bird list with your kids; in the process, you'll get to know the playground's avian residents and more easily spot new species. Make a countdown calendar to spring or fall migration to build your kids' anticipation for more opportunities to add birds to the list.

GOOD TO KNOW: *A playground is in essence a type of "patch"—a place close to home where you can look for and observe birds often. Consider submitting your lists to eBird.org to get your kids involved in citizen science.*

Tree Swallow

White Ibis forage on a Florida course.

59

PLAY SOME GOLF

GREAT FOR:

- Herons, egrets, and ducks in the course's water hazards
- Cranes, ibis, and other birds foraging on the greens, especially at or before first tee time
- Raptors perched on trees or posts lining the golf course perimeter
- Aerial feeders including swallows, swifts, and nightjars

GOLF COURSES ARE BIRD MAGNETS, but you often have to play golf to enter the habitat. There are some courses viewable from sidewalks and paths—just watch your head (especially if *I* ever take up the sport). My first bird-finding experience on the greens was with my dad years ago at Plantation Preserve Golf Course in Fort Lauderdale, which has a public walking trail packed with bird habitat that bisects the course. We found many birds in the microhabitats that peppered the greens: native trees and shrubs, marsh, fresh water. But I never expected what came next. On a small path close to the homes surrounding the golf course, our eyes were drawn to a flock of White Ibis foraging for mollusks in a shallow puddle on the lawn. There, alongside the ibis, was a bird I'd always wanted to see: a Limpkin.

If you play golf, consider booking the first or last tee time of the day to find even more birds as they forage on the grass. In Florida, I often see Cattle Egrets and White Ibis picking insects off the greens before golfers arrive. Any time of day, scan along the edges (see tip #23) where the greens meet trees, shrubs, marsh, or water hazards. And don't forget to stretch your neck after that killer putt; a golfer I know once spotted over a dozen Common Nighthawks buzzing overhead in the middle of the afternoon.

Limpkin

Brown Thrasher shows up for the author's yoga session.

60

DO SOME YOGA

GREAT FOR:

- Getting in tune with the natural world that surrounds you
- Putting nearby birds at ease so they'll be more likely to approach you

I DON'T KNOW HOW I ever lived without yoga. In some ways, the birds brought me to it: After countless hours of pounding-the-pavement urban birding, my muscles and joints were screaming for some love and care. For the past few years, a gentle twenty-minute session has been part of my morning routine. At home, I practice it facing the windows and often detect birds flying by: crows are cawing, House Sparrows are chattering, and Herring Gulls are gliding over the East River. Of course, the best yoga birding happens outside.

Whenever I travel, upon arrival I scope out a place to practice. At a house in Cape Coral, Florida, a screened-in lanai was where I laid down my make-shift mat (a beach towel) at sunrise. Mourning Doves cooed and provided the perfect yoga soundtrack, White Ibis and Boat-tailed Grackles foraged for tiny frogs on the lawn, and Great Egrets flew gracefully across the horizon. (A Northern Mockingbird tried hard to break my focus with its disjointed repertoire but did not succeed.)

Doing yoga outside helps you find more birds in several ways. You'll have to stay in one spot for a while (see tip #4) and the surrounding birds will get comfortable with your presence. If you're do-ing seated poses, birds may approach and forage at your side. In many poses, your gaze will be fixed on the horizon, which will make it easy to see birds in flight. Enjoy any birds that appear during your yoga session, but play it safe during challenging poses and maintain your gaze instead of following the bird with your eyes. At the end of your session, you might just want to hang out in Sukhasana (easy seated) pose and look for more birds!

GOOD TO KNOW: *Be careful around nests, especially those of aggressive defenders like mockingbirds! I met a woman who, while doing the downward dog pose one spring, was attacked by a Brooklyn mockingbird.*

61 WATCH YOUR STEP (ON THE BEACH)

GREAT FOR:

- Nesting and young plovers, sandpipers, terns, and skimmers
- Avoiding accidental damage to nests and eggs

NO JOGGING ON THE BEACH? That's what the sign says at the entrance to Breezy Point in Queens, New York. Each year from mid-March through August you can't run, fly a kite, play frisbee, or walk a dog. Similar rules, some more stringent than others, exist along many coastal areas of the US. Once you set foot on the beach, you'll not only understand these surprising restrictions, you will have found more birds. In roped-off, protected nesting areas, plovers, terns, oystercatchers, and other shorebirds sit on their nests, mere scrapes in sand or pebbles that are highly vulnerable to the human foot and a host of other threats. But the birds don't always stay within the ropes; their eggs blend right into the beautiful beach landscapes of shells, driftwood, and, in the case of New York City, trash. If you scan the beach a bit up from the shore above the tide line in mid- to late summer, after many of the chicks have fledged their nests, you'll often see tiny plovers scurrying through the sand with their even tinier—and fluffier—chicks trailing behind. Larger terns and oystercatchers will be easier to spot. It's best to take a seat on the sand and enjoy the show from afar, as even walking by will stress these already overworked creatures, busy teaching their young to find food, seek shelter, and follow their own rules of the beach.

GOOD TO KNOW: *During the COVID-19 pandemic, when beaches were empty, many shorebirds expanded their nesting zones, including the endangered western Snowy Plover in California.*

Black Skimmer chicks

COUNT VULTURES ON A ROAD TRIP

GREAT FOR:

• Kids on a road trip

• Finding hawks, eagles, falcons . . . and vultures!

• Spotting flying flocks

AFTER SPOTTING OVER A DOZEN Turkey Vultures while driving from South Carolina to Georgia, we decided to count them for the remaining two hours of the trip. I was shocked when our count quickly skyrocketed to 104. What's more, the trip flew by thanks to our entertaining high-count game with all passengers playing. Turkey Vultures are easy to spot on road trips since they circle and soar high above roads and open fields. Look for large birds that teeter while holding their wings in a shallow V shape (dihedral), especially after midday as heat rising off the roadways creates columns of thermals the vultures ride. Black Vultures can also be found on road trips. There are several ways to distinguish them from Turkey Vultures in flight, even at a distance: They hold their wings flat in line with their bodies, flap their wings more often, and don't teeter.

GOOD TO KNOW: *If you're road-tripping to the Florida Everglades anytime soon, you can enjoy close encounters with Black Vultures in parking lots—but beware. You'll notice something is awry when you pull in and see plastic bags hanging from car windows and wrapped around windshield wipers. These are meant to prevent the vultures from tearing off rubber and vinyl from cars. Certain parts of Everglades National Park offer tarps and bungee cords free of charge, but if that's not the case, you can protect your vehicle by covering vulnerable areas with plastic bags or wet sheets or towels.*

Turkey Vulture

Northern Mockingbird eyes the runways at JAX.

63

SCAN THE AIRPORT RUNWAYS

GREAT FOR:

- Raptors and vultures
- Local specialty birds
- Preflight and tarmac entertainment

AFTER AN INTENSE BIRDING TRIP to the Rio Grande Valley, with very early mornings and little sleep, Connor was relieved to get to the airport and rest his eyes from a week of looking for birds. But I wasn't finished yet. I had seen a photo of a Long-billed Curlew recently taken at the airport and had grand plans to look for it as we awaited our flight. I had never been more determined to find a bird. Once at the gate, I scanned the grassy areas beside the runways for large birds with long and downcurved beaks. It took only a few minutes to find a group of birds that possibly fit the bill, but I needed more magnification. Luckily, we had a spotting scope with us—did they allow that at the airport? Best not to ask. We acted quickly and set it up. And there they were—a flock of Long-billed Curlews foraging on the grass. As they took off in flight, they were easier to count: seventeen in all.

People are always finding cool birds at airports, which offer welcoming habitats such as grassy, scrubby open areas, retention ponds, and towers. At LAX, people have reported a Ferruginous Hawk hovering near a runway, a Say's Phoebe flying over a runway apron (plane parking lot), and a Burrowing Owl perched on a low parking lot fence. Passengers at Anchorage International Airport have had a host of interesting sightings, including a Snow Bunting on the runway, a Wilson's Warbler in a shrub by the gate, and even a Sandhill Crane flying straight down the runway.

Airport birding is especially exciting in birdy areas of the world where you're likely to spot a species you've never seen before. Be on the lookout from the tarmac as well; you may pass close to various grassy or water habitats.

GOOD TO KNOW: *Many birds are so drawn to the large open habitats around runways that airports have to take measures to keep these areas clear for takeoff. Some employ "airport dogs" to chase away birds and other wildlife, but not before giving them special training and noise-protective earmuffs.*

Snowy Egret

64

LOOK FOR WILD BIRDS AT THE ZOO

IF YOU'VE HEARD THAT "zoo birds" don't count toward your life list, you've only heard half the story. Visit any zoo and there will be wild birds to find; they visit or live on the grounds and take advantage of the food, shelter, and prime nesting sites that zoos have to offer. In fact, there are dozens of zoos that are eBird hotspots, including the National Zoological Park in Washington, DC; Jacksonville Zoo & Gardens; the Bronx Zoo; and the San Diego Zoo. At the menagerie in the Jardin des Plantes in Paris, I found several birds to add to my life list, including Carrion Crow and Rose-ringed Parakeet—a bird that sounds like it would be a zoo resident but is now wild in France, having successfully bred there since it was introduced in the nineteenth century.

Finding wild birds at zoos begins before you enter; check the trees and grounds near the ticket booths and listen for sounds. Common Grackles will noisily greet you from the top of the palms at the entrance to Audubon Zoo in New Orleans, Louisiana; Brewer's Blackbirds are often seen at the entrance to the Los Angeles Zoo. Once you get inside, most wild birds will be easy to spot as they freely forage the grounds, but wild waterbirds can be more challenging to distinguish. Getting to know the common ducks, herons, and egrets of your area using eBird or Merlin will help, and you might try asking zoo staff as well. Take advantage of raised platforms and viewing towers at zoos to scan the sky and trees; a tall platform overlooking the gorilla compound at Gulf Breeze Zoo in Florida afforded me an incredible view of a Great Blue Heron feeding its chicks at the very top of a 100-foot (30 m) pine tree. The scene looked so prehistoric that for a moment I thought I was in Jurassic Park.

GREAT FOR:

- Resident wild birds of the area, especially blackbirds, sparrows, woodpeckers, and wrens
- Nesting herons, egrets, and night-herons, including rookeries and colonies (see tip #81)
- Hummingbirds that feed on nectar at zoo flower beds

GOOD TO KNOW: *Many zoos, including the Smithsonian's National Zoo, monitor visits by wild birds.*

65

VISIT A CEMETERY

GREAT FOR:

Migrating birds,
woodpeckers,
raptors, owls

IN THE EARLY NINETEENTH CENTURY, before there were many public parks in the US, people flocked to "rural" cemeteries for outdoor recreation. Mount Auburn Cemetery in Cambridge, Massachusetts, which opened in 1831, was the first such cemetery in the country. Inspired by Père-Lachaise Cemetery (1804) in Paris, it was designed as a public outdoor space with beautifully landscaped gardens, trees, ponds, and lakes. The dearly departed rest there in crypts, graves, and memorials located among charming, winding paths.

Cemeteries offer habitats that attract both migrating and resident birds, including hawks and falcons that perch around the perimeter of open areas to hunt. The tranquil, uncrowded setting of cemeteries makes it easy to spot birds' movements and sounds. On a visit to my grandmother's grave in Connecticut, I met some of her cemetery companions: a Northern Flicker, a Black-capped Chickadee, and a Tufted Titmouse. One of my most otherworldly cemetery bird sightings took place while leading a small group on a walk in Brooklyn's First Calvary Cemetery; an American Kestrel appeared angelic as it hovered just above us and the tall memorial statues and graves.

The next time you visit the grave of a loved one, make a day of it and get to know the birds of their neighborhood.

GOOD TO KNOW: *Some cemeteries are planting more native trees, shrubs, and grasses and have reduced lawn-mowing schedules to lessen carbon emissions and noise pollution, attracting more native birds and pollinators. (The Green-Wood Cemetery in Brooklyn is one example.)*

*Eastern Bluebird at
Green-Wood Cemetery*

66

TAKE THE TRAIN

GREAT FOR:

- Soaring and perching birds of prey
- Foraging or flying flocks of large birds like cranes, cormorants, geese, and pelicans
- Close views of herons and egrets at the water's edge

BELIEVE IT OR NOT, years ago I traveled with a circus band, playing bass guitar in large arenas across the US. Probably the coolest thing about that was living on a train, in a little 7-by-7-foot (2 × 2 m) room with a surprisingly enormous window. The only problem was, I had not yet discovered the awesomeness of birds. Traveling through the country on beautiful back routes and being stuck in the middle of nowhere for hours would have made for some unbelievable bird-finding. Trains often pass through areas that cars and buses do not, and many routes follow a relatively straight path along rivers, open fields, and mountains for prolonged periods, making it easy to scan the adjacent landscape and sky for birds. The perfect onboard entertainment!

These days, I'm lucky to live near the Metro-North Hudson Line, which runs along the river of the same name. On a recent chilly day, I boarded the train and headed just thirty minutes north of the city in search of Bald Eagles. Within minutes of disembarking, I scanned with my binoculars and found two Bald Eagles circling low over the Hudson. Flying fast, one headed my way and passed less than 20 feet (6 m) overhead. On trips farther north, it's easy to spot birds from the train; I enjoy watching for Red-tailed Hawks and Great Blue Herons that perch on snags that overlook the river, as well as ducks and gulls in the water.

Amtrak offers many picturesque routes through the Rockies, the Grand Canyon, the Chihuahuan Desert, and the boreal forest of Canada, to name a few. Whether you're traveling near or far, the next time you hop on a train, grab a window seat and see what birds you can find.

*Bald Eagle soars in the snow at
the train station in Peekskill, NY.*

67

TAKE A BIKE RIDE

BIRDWATCHING BY BIKE is a thing, and many swear by it. With a medium pace that allows for comfortable listening and scanning, biking lets you find more birds in a short amount of time while maximizing your experience of the landscape. And biking has one of the lowest carbon footprints of any mode of transport.

During the holidays, we rented bikes on Hilton Head Island and passed by a tiny marsh with several ducks—comical-looking Hooded Mergansers with head feathers fully raised to attract a solo female. A bit farther down the bike path, a large bird darted by and started drumming in a tree; we didn't get a good look but suspected a Pileated Woodpecker, and verified it by checking the Merlin Bird ID app (see tip #45).

Seek out bike paths that run along water sources—rivers, reservoirs, levees, even neighborhood bird baths—for the easiest bird-finding by bike. Many suburban bike paths run through amazing wetland habitats.

Hooded Merganser

68

GO CAMPING

GREAT FOR:

• Crows and other local campground residents

• Sharing birds' outdoor habitats and more easily witnessing their interesting behaviors

WHEN I WAS A KID, my family never camped. As an adult, I finally planned a trip when I realized camping was one of the best ways to find more birds, namely, the elusive Black Rail. This tiny sparrow-size rail is so difficult to find that even some of the best birders in the world have never seen it, only heard it. To maximize our chances of finding it, we'd need time and patience. Basically, we'd need to hang around the bird's habitat for a while—what better way than to camp there?

As we pulled up to the campground, we rolled down the windows to listen before making the turn into the entrance. That was easy—well, to hear them at least. Black Rails were calling *kickee-doo, kickee-doo* from the pickleweed patch below. As we continued, we took care to avoid the bright Western Bluebirds flying down to the middle of the road to catch insects. Around the bend, a pair of Acorn Woodpeckers had taken up residence in a gigantic dead tree that now resembled a work of driftwood art, complete with cavities and embedded acorns that the diligent birds had stored. We marveled as we drove around it to enter the parking lot, where we were greeted by a family of Wild Turkeys and their awkwardly elegant chicks.

Check campgrounds' websites for information on common species found there; many campgrounds are eBird hotspots (see tip #52) and some even have stocked fishing ponds that attract cool aquatic birds like Anhingas and cormorants. In case you're wondering, I never did see a Black Rail at the campground, though Connor did. Luckily, we later caught a millisecond glimpse of one together at a nearby sewage treatment plant.

Anhinga takes advantage of a campground's stocked fishing pond.

Long-tailed Duck

69

TAKE A FERRY RIDE

GREAT FOR:

- Seabirds, some only visible out on the open ocean
- Diving ducks and other ducks not often seen from shore
- Gulls and terns

A PELAGIC BIRDING TRIP is a one- or multi-day boat excursion in search of birds that spend most of their lives in flight far out at sea. Shearwaters, petrels, and albatross are a few of these species that rarely touch land except to nest and breed, often on rocky cliffs. I tried and tried to venture out on a pelagic, but each time I booked one it would get canceled due to inclement weather, which is often the case with these trips. I once received a cancellation email that read, "We are sorry to have to cancel but don't want to run the trip from hell," and it was at that point I decided maybe the uncertainty of pelagic trips was not for me. But a few years later when my coworkers invited me, I couldn't resist—and we actually set sail! Large swells persisted for the first six hours (the trip was twelve hours in total), and quite a few people got seasick. I wore acupressure wristbands, which seem to have saved me. As we engaged our core muscles to stay planted on the swaying deck, we enjoyed views of Dovekies, Northern Gannets, and even a few Atlantic Puffins.

The good news is you don't have to book a pelagic trip to find cool seabirds; you can often spot them during a vacation cruise, a whale watching trip, or even a local ferry ride. On a Boston whale watching trip to Stellwagen Bank National Marine Sanctuary, I saw Wilson's Storm-Petrel and Great Shearwater for the first time, but it was the humpback whales that stole the show—a pair did a synchronized 360-degree breach less than 150 feet (46 m) from the boat!

Long-tailed Ducks are not considered true "pelagic" birds, but they're still awesome. They breed in the high Arctic and currently hold the prize for the deepest dives in the duck world, having reached depths of over 200 feet (61 m). Each winter I take the NYC Ferry out toward Rockaway Beach and marvel at dozens flying beside the boat. They're so ornate and entertaining that I barely notice I'm the only person riding outside due to the freezing temperatures.

GOOD TO KNOW: *If you want to see puffins (who doesn't?), you can take leisurely cruises in Maine to observe the extreme cuteness of Atlantic Puffins.*

FINDING BIRDS

through

THE COMMUNITY

JOIN AN AUDUBON SOCIETY CHAPTER OR BIRD CLUB NEAR YOU

GREAT FOR:

- Making bird-finding friends
- Getting help finding and identifying birds
- Learning the resident and migrating birds of your local area
- Finding out the ultra-local birding hotspots, especially birdy "microhabitats" within a park or preserve

I DON'T KNOW WHERE I'D BE today had I not joined my local Audubon chapter, the Francis M. Weston Audubon Society of Pensacola, Florida, back in 2010. There, I not only learned how to find more birds; I learned techniques for identifying them, how to use binoculars, the importance of habitat conservation and the many threats birds face, how to contribute to citizen science, and more. I also never would have spied a Florida manatee or eaten the largest flounder I had ever seen, both on an unforgettable Audubon field trip to Wakulla Springs. But most important, I made wonderful friends that I continue to learn from and grow with in birdwatching and life.

There are over 450 Audubon chapters in the US, so there's a good chance you can find one near you. They offer field trips to local, regional, and even global hotspots, where you can absorb the deep knowledge and experience of expert birdwatchers and become a part of the greater birding community. You can also attend interesting presentations and get involved in conservation efforts and programs. Visit audubon.org to find your nearest chapter. There are also many local and regional bird clubs and organizations that offer similar experiences.

Black Skimmer, the mascot of the
Frances M. Weston Audubon Society

ASK PEOPLE WITH BINOCULARS

GREAT FOR:

- Life birds, rare birds, local specialty birds
- Finding out the birdiest spots in a location
- Making bird-finding friends

YOU'RE ALMOST GUARANTEED to find more birds if you ask someone with binoculars around their neck what they've been seeing. In fact, spotting a binocular-clad human often means you're in a birding hotspot! It's especially useful when you're looking for birds in an unfamiliar place. When traveling, I always breathe a sigh of relief when running into another birder whom I can ask for the local bird report. Many have graciously pointed out a bird I had never seen before perched close by, like this Anna's Hummingbird in Arizona, or shared super-helpful information about the birdiest places in a park or preserve.

There are different responses and vibes you might get when you ask, "What are you seeing?" or "Anything good today?"—many friendly, others unfriendly—but none of that matters. Just ask, and you'll find more birds! Well, almost. On a trip to the Rio Grande Valley in Texas, I offended one man by asking, "What have you seen today?" He responded with annoyance, "What do you mean? Do you mean birds? I'm not looking at birds, I'm looking at butterflies!" Yikes, who knew such an innocent question could elicit such intensity? I apologized and asked what butterflies he was seeing, and he graciously pointed out some beauties. Lesson learned. When in Texas and other butterfly hotspots, you might consider first asking someone with binoculars, "Excuse me, are you looking at birds?"

Anna's Hummingbird

A tiny Pacific-slope Flycatcher in San Jose, CA. You see it, but your friends don't! Help them find it with clear directional cues and reference points.

72

GET YOUR FRIEND ON THE BIRD

GREAT FOR:

- Reducing frustration when trying to describe the location of a bird

- Helping friends and family find more birds

- Sharpening your ability to quickly detect birds among various landscapes

IF YOU'VE EVER BIRDED with a buddy, you know how surprisingly difficult it can be to communicate the exact location of a bird you're looking at, especially in a dense forest, an urban landscape, or when a bird is partially hidden behind a bunch of leaves. This can be frustrating, especially when you have found a rare bird that you desperately want to get your friend on while it's still in your sight! Learning this skill will not only help your friends see more birds, it will improve your bird-finding abilities. You'll start to notice yourself visually dividing the landscape into sections, scanning more systematically, and finding many birds you would previously have missed. But quickly getting someone "on a bird" takes practice, so where do you start?

With the "clockface" method, you can use a reference point like a tree, shrub, or water body. It's just like it sounds: Envision a tree in front of you as the face of a clock, such that the top of the tree is twelve o'clock and the bottom is six o'clock. Starting from the top, move clockwise to get the "hour" of where the bird is. This works great for trees with circular, oval, and triangular shapes when they aren't crammed together in a stand. Combine the "hour" with a mention of the distance from a tree's trunk or edge, and your friend will be amazed at how quickly they find the bird (for example, "The cuckoo is at four o'clock, a foot in from the outer edge of the tree"). To use this technique for birds on water (or other wide-open spaces), define an invisible line moving out into the water as twelve o'clock (from the bow of a boat, let's say), such that the imaginary hour hand will move along the plane of the water's surface.

Landmarks and lines are also super useful; think of the view in front of you as a geometric group of interconnected shapes and paths. Find something that stands out within the surrounding landscape: a utility pole, a sign on a building, another tree or plant, even another perched bird. Something far away works fine for this reference point. If possible, position yourself so that scanning down or across from the object in a straight line leads directly to the bird, and have your friend do the same. If a direct path won't work, connect various shapes—landmarks or other objects—using straight lines, so that your friend can easily follow the path to the bird. If the bird is in motion, adding an action clue to the end of your directions is also helpful. ("Starting from the base of this holly tree, scan right until you see the piece of trash on the lawn. The Palm Warbler is a foot behind it, pumping its tail.")

Birding with a buddy is much more enjoyable when you can share more birds!

73

ORGANIZE A BIRD WALK FOR YOUR COMMUNITY

GREAT FOR:

- Raising awareness of local birds
- Getting people in your community to care more about birds and the preservation of habitats
- Extra eyes to help each other find more birds

ONCE YOU'VE GOTTEN TO KNOW some of the usual bird suspects in your area, you now know more about local birds than most of the people who live there. Why not organize a bird walk? You'll not only have extra eyes to help you scan the skies and trees and leaves, you'll be raising awareness of birds in your local green space and getting people to care more about nature.

Sure, there are expert bird guides, but there are also hosts of others with varying levels of expertise. Offering a casual bird walk for beginners or a community outing to find birds is something we are all qualified to organize and lead. I could barely identify a mockingbird when I led my first walk years ago. I was nervous, but I got through it. Best of all, someone on the walk pointed out a tiny bird scaling a tree trunk and asked what it was. I never expected it to be a bird I had been trying to find for over a year—a Brown Creeper!

There are countless other exciting bird moments I would have missed had I not been leading a walk—a tiny Ruby-throated Hummingbird feeding on nectar high atop a Paulownia tree, a juvenile Eastern Meadowlark obscured amid some tall grass, a Black-crowned Night-Heron roosting in a London plane tree—all pointed out by other awesome bird-spotters. Organize a bird walk so you can help each other find more birds while having a positive impact on your community.

European Robins are a common sighting in London, UK,
and are easy to share with others on a bird walk.

Andean Cock-of-the-rock in Colombia

HIRE A BIRD GUIDE

- Safe bird-finding in unfamiliar locations
- Building your "life list," maximizing how many birds you find when visiting somewhere
- Letting someone else take care of the planning and driving!

JUST LIKE YOU ARE NOW becoming a bird-finding expert of your street, neighborhood, or local park, bird-lovers around the world have done the same and made a job of it. Professional bird guides have one of the coolest and possibly most difficult jobs on the planet—to help us find more birds! Most offer both group and private tours, and the latter can be customized to you.

The first time I hired a guide was on a trip to Colombia. I was not about to drive through the enormous landscapes of the Andes looking for birds on my own. A guide was a necessity to get me safely and securely to where the birds were. A bonus of hiring a guide is that they often have relationships with private landowners and nature preserves and can easily arrange visits at the best times to guarantee sightings of birds you've always wanted to see (your "target birds"). For me, that was the Andean Cock-of-the-rock. I had no idea how close we would get to these rock-star birds until our guide led us into the preserve—they were everywhere! One perched on a hanging flowerpot above our heads, others were sparring on the ground, and many were making a racket doing their courtship display low in the trees. We were smack-dab in the middle of a cock-of-the-rock "lek"—a place where male birds gather to put on impressive (and loud) courtship displays to attract females.

It can be challenging to find bird guides on the internet, especially since the term "bird guide" also refers to a type of book. Search for "guided birding tours" or "private birding tours" along with your desired city to get some good results. My favorite way is to ask people in my local birding community for recommendations of good guides, which is how I found my guide for Colombia.

GOOD TO KNOW: *In spring, several endangered and threatened grouse species gather on grassy plains of the US to dance; make deep, percussive booming sounds; and go wild for the females. These leks are sensitive areas where birds should not be disturbed; you can arrange a tour to view the show behind bird blinds. Some locations also take bird-blind reservations; all have strict viewing guidelines to protect the performers.*

GET TO KNOW PARK OR GROUNDS STAFF

GREAT FOR:

- Rarities, raptors
- Finding nests

IF YOU'RE PASSING THROUGH a park or other managed landscape, you have a great resource for finding more birds: the people who work there every day. Gardeners know the regular roosts of raptors, or which birds have been spotted recently in the park, and where. Maintenance staff know the nooks and crannies of the area and are experts at noticing something different (for example, a colorful bird or a giant raptor).

This bird-finding tip will work even better with a place you visit regularly—your birding "patch" (see tip #18)—because you'll develop a dialogue with the staff; it will become routine for you to exchange interesting bird and nature sightings. It can take a while to develop such relationships, but over time you will depend on each other for this information exchange. It will not only help you find more birds, it will also help the staff better understand how their work—planting, pruning, construction, trash management, and the like—might affect bird numbers and diversity.

That said, take the exact species names mentioned by park staff with a grain of salt. In search of the elusive Canyon Wren in one Arizona desert park, I inquired at the front office, where a ranger told me these wrens were "everywhere." I leaped out of the office with high hopes until I realized the ranger had thought the Canyon Wren was the area's ubiquitous Cactus Wren.

Cactus Wren

PARTICIPATE IN GLOBAL BIG DAY IN MAY

GREAT FOR:

- Contributing to citizen science
- Experiencing spring migration
- Meeting other birders

EACH MAY, at the height of spring migration, the Cornell Lab of Ornithology puts on Global Big Day—when people all over the world collectively find as many birds as they can in twenty-four hours. In 2022, the global bird-finding community found more than 7,700 species of birds—over 70 percent of the world's species—in a single day! (Birders in Colombia, the country with the highest bird biodiversity in the world, found 1,556 bird species during the event.)

Being a part of this global citizen science effort is highly motivating and will help you find more birds. Whether you're scouring the sand, the sky, or the cement, you'll be employing all of your bird-finding skills to contribute your sightings to eBird, one of the largest biodiversity-related citizen science projects in the world. It's easy to be a part: Simply sign up for a free account on eBird.org and enter one or more checklists of birds you see on Global Big Day. Joining a guided walk is also an option in many locations; check social media and local bird and nature organization websites for guided Big Day walks and events.

Mississippi Kites photographed during Global Big Day scouting in Florida

ASK THE LOCALS

GREAT FOR:

- Local news-making bird celebrities like owls and nesting hawks and eagles

- Specialty resident birds like established colonies of parakeets

- Getting inside information on where to find birds in the area

- Finding out the best places to eat!

PLENTY OF CITIES AND TOWNS have their local bird celebrities, from the resident Red-masked Parakeets of San Francisco's Telegraph Hill to a Snowy Owl that recently stopped by New York City's Central Park. Whether people are into birds or not, they hear news and community chatter about these cool birds in their neighborhood, the ones many of us are hoping to find. They're also usually quick to help once they see you wearing binoculars and may tip you off to the local hawk's nest (and the best lunch spots).

Following a lead from eBird's Species Maps (see tip #52), we headed to a golf course in the Florida Keys in search of Burrowing Owls—small, long-legged owls that are active during the day as they chill and hunt outside the burrows they use for nesting and roosting. When we arrived, we asked a local resident passing by if she knew about these curious creatures. Of course she did! She told us that, while the owls were no longer at the golf course, there were some in an empty lot in a nearby neighborhood, but she couldn't remember which one. Without hesitation, she gave us directions to the local real estate firm, saying they would surely know where to find the owls. This was turning into quite an adventure! It felt strange to ask the local real estate agent where to find more birds, but we decided to give it a shot. Surprisingly, we didn't get any strange looks at the agency; instead, the receptionist led us into the office of an agent, who grabbed his car keys and said, "Follow me." He led us to the lot and pointed to a marker sticking out of the ground. "That's the burrow," he said, and drove off as we thanked him profusely.

Burrowing Owl

FINDING
BIRDS

ACTING
CRAZY

FOLLOW THE CAW OF CROWS

GREAT FOR:

- Experiencing exciting bird drama, often involving raptors
- Watching birds go into territorial and nest defense mode against crows

IT TURNS OUT THAT CROWS are awesome at helping you find more birds. When you hear them getting noisy and rowdy, chances are good that a raptor is nearby. Hawks, falcons, owls, even eagles get "mobbed" by crows, who band together to drive these larger potential predators out of their territory. Mobbing is an intense aerial drama that's not to be missed, as the crows chase, dive at, and hit a raptor in flight. And when the birds fly out of sight behind a tree or building, that doesn't mean the show is over; they often fly back around a minute or two later and continue to entertain.

After taking a few cues from the crows, you'll soon realize how easy it is to find a bird of prey nearby. It will become instinct to find the bird being bullied. When I hear a crow in my patch, I immediately look to the sky for a soaring Red-tailed Hawk. The target species and circumstances may be different where you live, but if you have resident crows and raptors in your neighborhood, a certain mobbing scene will play out time and time again. In addition to Red-tailed Hawks, crows also mob Sharp-shinned and Cooper's Hawks, Great Horned and Barn Owls, Northern Goshawks, Merlins, American Kestrels, and more.

During spring and summer nesting season, hearing caws may also mean that a crow is the one getting mobbed—by smaller birds such as sparrows and swallows who have fallen victim to an attempted or successful nest robbery. (Crows are adept at stealing eggs and nestlings.)

American Crow mobs a Red-tailed Hawk.

79

FIND FERMENTED BERRIES

SOMETIMES BIRDS ACT SO CRAZY—flying erratically or passing out—that people call the police. The culprits are usually intoxicated Cedar Waxwings, flocking birds whose diet consists primarily of berries, that have ingested more ethanol than they can handle. (This also happens to American Robins, big berry-eaters in winter.) After a frost, berries start to ferment, and their ethanol content increases. What appears to be the ultimate banquet might actually be a very berry bar. Cedar Waxwings are notorious for eating voraciously; a fully fruited tree can easily be plucked bare by a flock in one day. And although they are well adapted to their berry diet, with short intestines that move food through quickly and large livers to process carbohydrates and filter toxins, eating that much fermented fruit can take its toll.

Luckily, many birds recover. But accidents do occur; flocks of birds hitting windows and cars are what often prompt calls to the police. If you find an injured imbiber, contact your local wildlife rehabilitation center for help.

GOOD TO KNOW: *If you're ever in the Caribbean and notice an open-air bar with liquor bottles wrapped in plastic, blame it on the Bananaquits. These fearless nectar-eaters visit beach bars for drinks and even steal sugar from occupied restaurant tables. They also pierce the base of flowers for a nectar snack, much like orioles in the US (see tip #40, "Look for fallen flowers").*

Cedar Waxwing

80 INVESTIGATE ANY AWKWARD FLYING OR COMMOTION

GREAT FOR:

- Bird drama including birds like crows or grackles robbing nests
- Territorial defense
- Finding nearby raptors

IN THE WORLD OF BIRDING, you never know what you might find. I could easily have missed this strangely cute punk-rock baby grackle had I not investigated some birds acting crazy. I was used to seeing Barn Swallows but had never seen them doing this—a dozen were flying constantly in a tight circle just off the edge of the pier they build their nests under. I had to look through my binoculars to locate their source of frustration, a Common Grackle perched on the pier's edge. Now it made sense; grackles are classic nest robbers, often snatching not only eggs but also nestlings for a high-protein snack. Still, it seemed odd; why hadn't the grackle taken off by now? (Grackles are usually not as persistent as crows in the same endeavor.) I decided to walk over to the sports courts directly on the pier to get a better view of the drama. And that's when I saw *this* thing: a Common Grackle fledgling in the most awkward stage of its plumage development. It was right next to the adult that was enduring the mobbing of Barn Swallows only because it was busy feeding this little guy that could not yet fly.

I took a seat nearby (see tip #9) and almost immediately, the punk rocker started running toward me. Its parent started to vocalize in an apparent attempt to draw its baby back, but it kept approaching me. I seriously think it thought I was its mother. Whatever the reason, I wouldn't trade anything for this photo. Always investigate bird commotion.

Common Grackle fledgling ready to take on the world

81

VISIT A ROOKERY

GREAT FOR:

- Seeing large egrets and herons
- Witnessing courtship, baby birds, and birds acting crazy

WHEN I LIVED ON PENSACOLA BEACH in Florida, each day I'd drive right past a tiny park near the beach tollbooth. Little did I know that during summer, this spot—just over half an acre—was a bird haven. Luckily, I happened to notice a large white bird acting crazy on top of an oak, flailing its wings and appearing quite prehistoric. It was time to check out this tiny park. I was met with a myriad of the craziest squawking sounds I had ever heard. The wooden platform running through the park was covered in whitewash—bird poop (see tip #39). The squawkers turned out to be Snowy Egrets; I looked up to find several with their elegant aigrettes (breeding plumes) billowing to and fro as they engaged in courtship. A cluster of tree-top nests caught my attention; I had found a rookery (a breeding site of a group of gregarious birds)!

Rookeries contain many large nests of herons, egrets, storks, and spoonbills clustered close together in trees over water or on an island. The best way to find one in your area is to search online for "rookery" and your state or city. Some are located within zoos; Lincoln Park Zoo in downtown Chicago, for example, is home to a Black-crowned Night-Heron rookery. And the St. Augustine Alligator Farm Zoological Park might be the noisiest place on earth in spring and summer when it becomes a rookery for various species, including Wood Storks, Roseate Spoonbills, Great Egrets, and more. These birds take advantage of the free alligator patrol below that prevents raccoons and other predators from reaching their nests to take eggs and chicks.

Great Egret

A rookery in Florida

Male Gadwall tries to attract nearby female.

82

STOP FOR DUCKS DANCING AND PIGEONS PIROUETTING

GREAT FOR:

Witnessing birds' fascinating (or funny) courtship displays and mating

IT'S EASY TO PASS BY BIRDS you see every day without realizing they're seconds away from engaging in fascinating courtship and mating rituals. I did this for years—I often noticed a duck bowing and thought it was cool but kept moving on. Little did I know this was a signal to stick around and watch the awesome show that is duck courtship and mating. In addition to bowing, look for waterbirds chasing each other or racing over the water. Loon pairs often do the latter in unison, racing atop the water as if onstage in a ballet. Despite appearing much like tiny toy ducks, male Bufflehead ducks are more aggressive—albeit hilarious—in their courtship; they chase other males away from their female interest while bobbing their heads forward and back.

One of the most entertaining and easy-to-see courtship rituals is the dance of the Rock Pigeon. The next time you spot a puffed-out pigeon at your feet that's following closely behind another, get ready. Watch as it bows repeatedly and makes full turns to impress the female. The show may end with copulation, but more often than not the female takes off in flight. Rock pigeons and other doves, including Mourning Doves, often exhibit a charming courtship that includes gently nibbling each other's face and neck (allopreening) and the female placing her bill in the male's (I call it "kissing" but ornithologists call it "billing"). Further cuteness and mating often follow both of these seemingly romantic gestures, so take a seat (see tip #9) when you encounter them.

Also watch for birds bearing gifts for their mates. Some present sticks, reeds, and other nesting materials to potential partners; watching a Double-crested Cormorant do this was one of the most beautiful things I've seen in my patch. Others gift sweet berry snacks, crunchy insect protein shots, or even shiny sardines. There is often some postur-ing, fluttering, or yelping involved; these Common Terns posed with bills to the air during their gift exchange.

Common Tern courts mate with sardine.

American Robin nestlings

83

CHECK BIRD DETERRENT SPIKES

GREAT FOR:

- Close-up view of nests of pigeons, House Sparrows, gulls, and more
- Funny photo ops!

ONE SURE CLUE that birds are around is the presence of bird deterrents such as spikes, or fake owls—even fake coyotes—meant to prevent birds from perching or foraging. But don't be fooled; these supposed deterrents often have the opposite effect. Many birds with crazy tiptoeing skills use anti-bird spikes as structural supports for their nests, or as "safety gates" to protect their young from nest-robbing crows, jays, and grackles—birds too large to land or swoop between the spikes.

It turns out that spike-nesting birds are quite an inspiration. When I searched on Twitter, I found many people cheering them on as the ultimate go-getters, role models for overcoming the odds. Apart from plenty of photos of pigeons and House Sparrows, there were Barn Swallows, Herring Gulls, Black-legged Kittiwakes, and European Robins nesting among the spikes. Crows had even been photographed using them as nesting material after pulling them off buildings.

Check spiky habitats on buildings, ledges, and under awnings for nests at the onset of spring and through summer. I found the American Robins pictured on the previous page nesting on the lower level of Brooklyn Bridge Park's pedestrian bridge. Their anti-bird spiked nest made it impossible for the neighborhood crows to snack on their eggs and chicks.

Canada Goose approaches a fake coyote meant to scare it away.

FINDING BIRDS

you've

ALWAYS
WANTED
TO SEE

84 WILD TURKEYS

IF YOU HAVEN'T YET met a Wild Turkey, you're in luck. These impressive birds can be found in all states except Alaska, as well as in parts of Mexico and Canada; their population is estimated to be close to seven million. It's hard to believe these birds were nearing extinction in the late 1930s, when overhunting and habitat loss had reduced their population to around thirty thousand. But in what is often touted as one of the most successful conservation efforts in history—a Wild Turkey relocation program that started in the 1950s—the birds made a huge comeback and continue to increase in number.

Turkeys like to forage in fields and clearings adjacent to woodlands, so if that sounds like your backyard, be on the lookout. You can also use the "Species Maps" feature at eBird.org/explore (see tip #52) to find out where turkeys hang out near you. When visiting, listen for their *gobble-gobble* sounds, especially in spring when the toms (males) are courting the hens (females). Be sure to look up near the source of the sound; males often make gobbling noises from trees. Also keep your ears open for clucking, cackling, and purring, and scan for turkeys foraging on the ground, especially along edges where wooded areas meet open areas, even parking lots.

GOOD TO KNOW: *Turkeys from the small remaining populations were relocated to other areas with suitable woodland habitats, which included many farmlands abandoned after the Great Depression that had started to regrow into their original forest habitat.*

Wild Turkey

American Robin

85 BABY BIRDS: FOLLOW BIRDS CARRYING SOMETHING IN THEIR BILL

IT'S ONLY A MATTER of time until you notice a bird carrying something in its bill. Sometimes it's food for themselves or their young in the form of berries, insects, or even another bird. But in spring and summer, you'll see birds carrying twigs, sticks, branches, grasses, mud, string, or strips of plastic. Whatever it is, keep your eyes (or binoculars) on the bird as long as you can. If you see where it lands, celebrate! You have found its nesting location and can expect several weeks of baby-bird cuteness and fight-to-the-death drama. (The

nesting pair will be diligently de-
fending their nest against jays,
crows, and grackles that rob eggs
and fledglings, often to feed their
own young.) Note that some baby
birds, such as ducks and shorebirds, will leave the nest within
a day or two of hatching; these
birds are precocial, fluffy with
down, and able to move and feed

Cedar Waxwing nestlings

at hatching. They'll likely stay in the general area and
their parents will continue to protect them as they
grow. Many birds whose nests you'll find are helpless
on hatching and must develop in the nest; they're al-
tricial, born naked with eyes shut, and fully depen-
dent on their parents for food, warmth, and shelter.

Be sure to observe a nest site from at least 20
feet (6 m) away, possibly more. Birds can abandon
nest-building if they feel that the site is no longer hidden and safe. And getting too close to a full nest puts
stress on the birds busy raising their young. A big clue
that you're too close is if a parent has food in their
mouth but is not feeding its young. Back up to watch
the bird place a juicy worm or insect into one of its
nestlings' gaping mouths.

GOOD TO KNOW: *Find even more baby birds by tuning in to the begging
cries, peeps, and screams of young birds inside the nest (nestlings) and out
(fledglings). These strange and curious sounds often fall under our radar,
since they're high-pitched and faint. Knowing when to listen for them is the
first step in finding the birds that make them; in North America, tuning in by
June is a good bet. Listen well into summer for more chances to see baby birds.*

Barn Swallow fledglings begging for a tasty insect meal

Red-tailed Hawk in New York City

86 | LEARN THE SCREAM OF THE RED-TAILED HAWK

KYEEERRRRRR! Knowing the blood-curdling scream of a Red-tailed Hawk can lead you to an entertaining avian drama. You've likely heard it before—the two- to three-second scream is used all over the place in film, TV, and video games, most often to replace the ironically wimpy cry of the Bald Eagle. Listen to it now on YouTube or at macaulaylibrary.org; repeat for five minutes so you'll never forget it.

A screaming Red-tail often means other birds are around, irritating the hawk or intruding on its territory. It could be another Red-tailed Hawk, a murder of crows, a Cooper's Hawk, Bald Eagle, or Great Horned Owl—just a few of the species that spar with Red-tails. When you hear the sound, look up—they usually scream while soaring. If you hear but don't see a hawk in the air, check posts, trees, and other perches. But if all you come up with is a Blue Jay or Steller's Jay, you may have been duped; these talented birds can imitate the scream of the Red-tailed Hawk (and those of other hawks and falcons). Listen closely to jays' mimicry so you can better detect their weaker, raspier versions.

Red-tailed Hawks are one of the most common hawks in North America. Even if they aren't screaming, you can find them by scanning the sky (see tip #8) and looking for a large, bulky raptor with a rusty-red tail or a streaked belly band (young birds don't sport the colorful tail) circling above. Birding from the backseat (see tip #54) is another great way to find Red-tails. Look for them perched on posts and lights along highways, where they often wait for fresh mammal roadkill.

GOOD TO KNOW: *Several theories exist as to why jays imitate the calls of Red-tailed Hawks and other raptors, the most popular being that it's a way for them to scare other songbirds away from bird feeders and other food resources so they get the food all to themselves.*

HUMMINGBIRDS: VISIT RED FLOWER BEDS

HUMMINGBIRD FEEDERS ARE USUALLY RED, and for good reason—the color is a hummingbird magnet. Hummers have a fourth cone in their retina that enables them to detect colors on the ultraviolet spectrum (our three cones only detect red, green, and blue light). Colors ranging from red to yellow stand out to them, and they associate red flowers with having large sources of nectar. Fake flowers dupe these beauties too; I was excited yet saddened to see a hummingbird visit red plastic blooms that adorned the canopy of a Brooklyn brunch spot. There are also plenty of stories of hummingbirds nearly "attacking" someone wearing a red shirt or red floral skirt!

So make a habit of checking red flowers, especially those with a tubular shape. On my summer visits to the Brooklyn Botanic Garden, I head straight for the cardinal flowers to find Ruby-throated Hummingbirds hovering at the blooms. After taking a few sips, they often fly up to a nearby tree to catch some shade until the next sugar craving strikes. While red is the most attractive color to hummingbirds, they visit flowers of many colors; in my birding patch, their favorite nectar sources are the long purple flutes of the Paulownia tree and dangling orange jewelweed blooms.

GOOD TO KNOW: *If you have a hummingbird feeder, you can forgo red-colored nectar. There is no evidence that hummingbirds prefer it and it's not yet known whether artificial or natural red dyes might harm them. As long as the feeder itself is red, it will be great for attracting hummingbirds. To up your backyard red factor without dyes, accent with red pottery, furniture, and of course, flowers!*

*Juvenile Ruby-throated Hummingbird
at cardinal flower*

Bald Eagle perches alongside river.

88 SEE A BALD EAGLE

IT'S HARD TO BELIEVE that when I was born, Bald Eagles were on the brink of extinction. Had it not been for the banning of DDT and for protections afforded by the Endangered Species Act, I might never have met this impressive eagle flying high over my local park. Bald Eagles can be found across the continental US and Canada, with thousands breeding in Alaska, British Columbia, and Minnesota.

To find Bald Eagles, stop by their favorite "restaurants"—coastal areas, lakes, rivers, canals, mangroves, and swamps that serve up plenty of fish. Winter is the best time to look, when the eagles leave the frozen waters of Alaska and Canada behind and gather in large groups along water bodies in the contiguous US. In spring and summer, check forested areas near

Juvenile and adult Bald Eagle

water where they'll be building nests, courting, and raising their young. They tend to perch high and prefer very tall trees, especially pines and other conifers when available. Otherwise, check oaks and aspens, and even mangroves in Florida. Any time of year, scanning the sky (see tip #8) can yield a Bald Eagle; this is how I usually find them in New York City, and this tip is highly recommended in all urban and suburban areas.

Use tip #52 for up-to-the-minute reports of Bald Eagles in your area: Check eBird "Species Maps" for recent sightings and eBird "Bar Charts" to know what time of year you can expect to find the most Bald Eagles.

GOOD TO KNOW: *If you see a large raptor without a white head, you may have found a Bald Eagle. It takes four to five years before their characteristic stately white head plumage grows in and fully glows. Juvenile Bald Eagles look similar to Golden Eagles (spotted more often in the western US), so snap a photo for Merlin to verify, or consult your field guide.*

89 WOODPECKERS: LISTEN FOR DRUMMING

WALKING BACK FROM the grocery store along a busy Brooklyn street, I heard loud, resonant drumming in the distance. Although it sounded like a woodpecker, I discounted it as wishful thinking and continued home. But the sound grew louder and louder; soon there was no doubt that it was a bird. I sped up as the sound drew me to its source, two blocks from where I had first heard it. The rich, woody beat was now emanating from the tree right in front of me—but I couldn't find the bird! I cranked my neck and scanned the branches of the tall oak with my naked eyes the best I could. Just as I was about to resign myself to the fact that the bird was hidden from my view, my gaze landed on the female Downy Woodpecker, drumming away at eye level just 3 feet (1 m) in front of me.

Both male and female woodpeckers drum on many objects—trees, houses, chimneys, windows, playground slides—whatever they can find in their environment that will amplify their message to potential mates and drive would-be intruders from their territory. Many woodpeckers drum year-round, but it's easier to tune in to their beats in winter and spring as mating season approaches. To prep for your woodpecker encounters, listen to the different drumming and rhythms of woodpecker species found in your area at macaulaylibrary.org.

GOOD TO KNOW: *The sight of small holes drilled into tree trunks can also clue you in to the presence of woodpeckers, especially sapsuckers (see tip #43).*

Downy Woodpecker

Barred Owls

90

OWLS

THE FIRST STEP IN FINDING AN OWL is believing that you will. Have faith! It can take a long time to come face-to-face with an owl, and it often seems like a far-off dream. But there are owls living near you, and it's definitely in the realm of possibility that you can find one.

Start by discovering the most common owl species in your area by visiting eBird.org/explore. There's a good chance it's the Great Horned Owl—the most common owl in North America—able to thrive in a variety of habitats including woodlands, wetlands, deserts—even urban parks. Get familiar with the sound of this owl at macaulaylibrary.org. (Listen to the similar-sounding Mourning Dove to learn the difference.) Great Horneds are usually most vocal within the first hour after sunset, but you might hear them at different points during the night and before sunrise. While sound is your best clue to their presence, you can also find them much as you do other raptors: by checking high in trees, on posts, and on the edges of buildings for their shapes and silhouettes. Barred Owls, found in the eastern US and the Pacific Northwest, are less conspicuous and usually perch in dense woodland during the day. Listen for their classic "Who cooks for you?" song (one of the few mnemonics that works for me) at night.

Great Horned Owl

Smaller owls you might meet include screech owls (check tree cavities) or Burrowing Owls (check open grasslands and fields for them standing beside their burrows). The smallest owl in the world is the Elf Owl, around 5 inches (13 cm) long and found in the southwestern US. I still can't believe I've seen it; on a tip from an Arizona birding lodge where we stayed, Connor and I waited in a parking lot well after dusk to witness the male Elf Owl fly to a tree cavity and present a tasty insect to its mate.

Investigating other birds can also lead you to an owl. On a nature trail in Florida, I heard a woodpecker drumming; when I turned my head to investigate the source of the sound, I saw not one but two large upright football shapes through the trees—a pair of Barred Owls. Mockingbirds, crows, and even smaller birds defend their territory by mobbing owls, so follow tip #80 and investigate any bird commotion or wild sounds.

GOOD TO KNOW: *eBird is a great resource for finding locations where owls have been sighted in your area. However, many owls are considered sensitive species, so to protect them from disturbance by birders and photographers or from captive trade or hunting, their current locations are not displayed. Visit eBird.org/ explore and view "Species Maps" to see where owls have been reported in the past.*

Burrowing Owls

91 SEE A STORK

A STORK IN THE US? I had no idea they were here until I started watching birds. When I found out, I just had to see a Wood Stork, so I planned a trip to find them in Florida using locations reported in eBird (see tip #52). As we drove eagerly toward the coast on Interstate 10, our bounty materialized earlier than expected; a Wood Stork was foraging in a grassy ditch on the side of the road. These large white birds with bald, blackish heads and bills currently breed in the Southeast—Florida, Georgia, and South Carolina—but are expanding their range northward and have recently shown up in the Northeast. Wood Storks usually feed by tactolocation, groping for food rather than looking for it; they're often seen with their heads down in the shallow water of ditches, retention ponds, swamps, and other wetlands.

The best way to guarantee you'll find a Wood Stork is to plan a trip to a known rookery, or nesting colony, in Florida, Georgia, or South Carolina. I like to visit the St. Augustine Alligator Farm Zoological Park in April or May, but there are many options, including Crooked River State Park in Georgia and the trail at the Port Royal Cypress Wetlands and Rookery in South Carolina. Search "visit Wood Stork rookery" for more options, and use eBird.org/explore (see tip #52) to find out the right time of the year to visit. (April and May are a good bet, though some storks nest during winter in Florida.)

GOOD TO KNOW: *When in Florida, scan nearby retention ponds and wetlands for the large, white, upright shapes of Wood Storks. In Louisiana, July and August are prime stork times, when they frequent shallow crawfish ponds to feed.*

Wood Stork

American White Pelican

92 PELICANS

FOR SOME REASON I have the image in my mind of pelicans being the ones to deliver babies in their bills, but that is the fairy-tale job of storks. It could be because pelicans have giant, expandable pouches on their throat, which expand to accommodate the many large fish they catch. Either way, there's no denying that pelicans are exciting to see. I think of them as modern-day pterodactyls, with their prehistoric-looking heads and bills. As they soar above, I imagine what it would be like to witness similar flying creatures from long ago.

Brown Pelican

There are two types of pelicans in North America: the American White Pelican, which resembles those often portrayed in cartoons, and the Brown Pelican, an impressive diver that plunges into the water from as high as 60 feet (18 m), protected from the forceful impact by special air sacs that cushion their dive. Brown Pelicans are a common year-round sight along the southern coasts of the US. Look for them, often in groups, making large splashes as they plunge for fish. They also fly parallel to shore just above the water and soar in flocks of a dozen or more over beach homes and sand dunes.

American White Pelicans are not as common and take a bit more work to find. They scoop fish up from the water's surface, so you won't see them diving. But during spring and fall migration they move in large flocks, often flying remarkably high. Make sure to look up (see tip #8) periodically; I've encountered flocks of a hundred in Florida just by quickly scanning the sky when getting in and out of the car. They hang out mostly along coasts in winter but move inland to rivers, lakes, and marshes during summer. For zoning in on exact locations of American White Pelicans, check eBird.org/explore (see tip #52).

Peregrine Falcon with prey

THE FASTEST ANIMAL ON THE PLANET

NINE TIMES OUT OF TEN, when I ask a child in New York City to name their favorite bird, the immediate response is, "Peregrine Falcon!" (The other times, of course, they say, "Pigeon!") They've surely heard that peregrines are the fastest animals on the planet—having been clocked at 242 mph (389 km/h) in a dive—and are found on all continents except Antarctica. What's more, New York City has the highest density of these birds anywhere in the world. An increasing number of big cities, including Chicago, Philadelphia, and London (with the second-highest population), have seen increases in Peregrine Falcon populations following the banning of DDT and successful reintroduction efforts in the 1980s and '90s.

In cities during spring and summer, look for peregrines perched on or flying around bridges, skyscrapers, churches, and towers, their favorite urban nesting sites, which allow them to easily scan for birds they can snatch to feed their young. (Also search local news stories and videos for reports of these newsmakers nesting near you.) Any time of year, check the tail end of a flying pigeon flock for the sleeker shape of a determined falcon pursuing its prey. Outside of the spring and summer breeding season, peregrines often hunt shorebirds and ducks in wetlands and other open areas.

GOOD TO KNOW: *Peregrine Falcons nest on the Houses of Parliament in London, where they often hunt Rose-ringed Parakeets—escaped exotic birds with bright green plumage and large red beaks. The tens of thousands of these bright birds in the city presumably make an easy catch for the swift peregrine.*

94 A LARGE PINK BIRD

I WISH I COULD tell you it was easy to find flamingos in Florida. *Plastic* flamingos are easy to spot here in Red Hook, Brooklyn, and I confess I even get a rush seeing them! But the gorgeous, living, breathing kind of flamingos are quite tough to come by in the wilds of the US. The good news is that there is talk of an American Flamingo comeback in Florida; sightings are slowly increasing due to recent improvements in the wetland habitat they need to thrive, with clean water and plenty of crustacean snacks. (The Comprehensive Everglades Restoration Plan/CERP is a federal and state partnership to clean and restore the Florida Everglades over more than thirty years, benefiting flamingos, spoonbills, and the entire Everglades ecosystem.) When in Florida, it's worth checking eBird.org/explore for recent reports of the American Flamingo as well as searching the local news.

In the meantime, it's much easier to find a different giant pink bird—the Roseate Spoonbill. They hang out in wetland habitats of the southeastern US, and they dip and then swing their bills in shallow water to scoop up small fish and crustaceans. At a glance, they are often mistaken for flamingos, but with a closer look, their spatulate bills and bald heads are unmistakable. You can find them in both saltwater and freshwater, even in marshes. I've had the most luck by checking the many retention ponds of South Florida, even those behind superstores or hospitals. It's worth a short detour around the block to scan such ponds for spoonbills, and you might also find Anhingas, Wood Storks, herons, and more.

Roseate Spoonbill

FINDING COOL BIRDS

 you

MIGHT NOT HAVE HEARD OF

!!!

95 KING OF THE PANTRY

MANY BIRDS CACHE THEIR FOOD, stocking up for when resources become scarce. These include various species of jays, chickadees, titmice, and nuthatches. While the ability of chickadees to recall where they have hidden thousands of snacks (mostly seeds, but also insects) is impressive, the Acorn Woodpecker puts on the ultimate caching show. These woodpeckers of the western US create "granaries" in trees and human-made structures, riddling them with holes that they fill with acorns—one per hole. The best part is that these birds are easy to find if you visit their preferred habitats, especially oak and pine-oak woodlands of California, Oregon, Washington, and the Southwest. They are generally year-round residents (only moving on when acorn supplies decrease), though they'll be easier to notice in late summer and fall as they work busily all day long to store food for winter. Listen for their rambunctious squawks and banging as they excavate their granaries and carefully place acorns in holes.

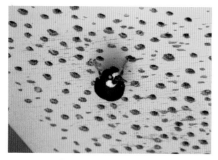

Acorn Woodpecker tending to its granary

At Guadalupe Oak Grove Park in San Jose, California, I encountered dozens of Acorn Woodpeckers in the oaks and the ground below them, but the loudest bird of the bunch was under the awning of the restroom building. I found it by investigating an extremely loud banging, which led me to the bird tending to its artsy awning granary.

Acorn Woodpecker

96

NORTH AMERICA'S SMALLEST FALCON

WHEN LOOKING FOR RAPTORS, it may seem intuitive to look for something large and stately. But the American Kestrel, the smallest and most common falcon in North America, is the size of a dove. It hunts insects and small mammals from above open areas, making it one of the easiest raptors to find—and the male's colorfully patterned plumage of slate blue and reddish orange makes it easy to identify. Check utility poles, light posts, fences, and wires for this medium-size, bulky, and big-headed bird. Kestrels often appear to teeter on a perch as if they're losing their balance; use this movement clue along with a bird's size and shape to detect them from far away. Also scan over fields for kestrels hovering in place as they gaze downward for prey.

Kestrels also hunt small birds, which seems to be their meal of choice in my local park. I often see them flying with a sparrow hanging from their tiny talons. When you see sparrows foraging on a lawn, check surrounding light posts and poles for a perched kestrel eyeing the flock. Then watch the lawn from a nearby path for a chance to see the falcon swoop down to snatch a sparrow snack. (If you wait on the lawn, the kestrel won't be as likely to dive in.) Track it as it flies to its perch for a chance to witness a "feather snowstorm" as the bird plucks its prey. (See tip #44.)

GOOD TO KNOW: *American Kestrels are "secondary cavity-nesters," nesting in existing cavities excavated by woodpeckers or in other hospitable holes. Holes in the cornices of Brooklyn brownstones are one of their urban cavity choices.*

American Kestrel perches on Brooklyn-Queens Expressway exit sign.

Vaux's Swifts enter chimney at dusk.

97 SWIFTS: HANG OUT AROUND CHIMNEYS IN SUMMER

SOMETHING FASCINATING IS HAPPENING inside many chimneys of North America during summer; small birds called swifts are nesting and roosting there. Following tracks (see tip #42) won't work for these birds; they have pamprodactyl feet—all four toes facing forward—preventing them from walking or even perching. This is why they frequent chimneys, where they can rest by clinging to the ridges between the bricks with their tiny feet. Using their super-sticky saliva, they glue twigs to the bricks to form half-cup nests.

When you're near chimneys in summer, listen for frantic chatter or twittering above as swifts hunt insects to eat or bring to their young. (You can listen to recordings of Chimney Swifts and Vaux's Swifts at macaulaylibrary.org.) When you hear them, look to the sky for small birds that look like bats, quickly fluttering their wings. (The renowned ornithologist Roger Tory Peterson described them perfectly as "cigars with wings.") An hour before dusk, look toward chimney entrances and you might catch anywhere from one to tens of thousands of swifts entering to roost for the night.

Chimney Swift

September is the best time to catch the spectacle of large flocks of swifts entering chimneys at dusk. Along the Pacific flyway in Washington, Oregon, and California, thousands of Vaux's Swifts may enter a single chimney to roost for the night as they migrate south to Mexico, Guatemala, and Honduras. In the eastern half of the US, Chimney Swifts do the same in smaller numbers as they head to their wintering locations as far south as Peru.

GOOD TO KNOW: *Chimney and Vaux's Swifts used to roost and nest more often in hollow trees or snags; as that habitat declined, they adapted to something similar: chimneys. Nowadays, chimneys are also in decline, but constructing swift towers can help. Plans to build such structures can be found for free on the internet.*

Loggerhead Shrike eats lizard it has impaled.

98

SHRIKES, AKA "BUTCHER BIRDS"

SHRIKES MIGHT JUST BE the fiercest birds in North America. They're like mini raptors, complete with sharp, hooked bills for hunting mice, lizards, large insects . . . even birds. Like kestrel falcons, they hover in place while scanning for and targeting prey. But there's one thing they're missing: talons to tear apart or otherwise prepare some of their larger meals. So what do they do? When they snatch a lizard or other vertebrate, they paralyze and kill it by biting its nape in just the right spot. (Their insect food prep is not as gruesome; they chomp on an insect's thorax to subdue it.) They then impale their prey on a nearby barbed wire fence, thin tree branch, or sharp-edged plant to finally enjoy their meal. There's a good reason shrikes are often referred to as "butcher birds."

Loggerhead Shrike scans for prey.

There are two species of shrike in North America: Northern Shrikes—which are tough to find, as they breed in remote locations in Alaska and Canada and only visit the northern portions of the continental US during winter—and Loggerhead Shrikes, which are easy to find in the southern US, where they are often year-round residents. Loggerheads are usually found solo, hunting from or resting on exposed perches like power lines, fences, even park benches and mailboxes. Though they're small, their chunky silhouettes stand out even from a distance. If you notice such a shape, watch for patches of white on the bird's wings as it flies. (The larger, medium-size Northern Mockingbird also flashes its white wings in flight.) I have found that you can get pretty close to shrikes without scaring them off unless they have a nearby nest or they're with their young. Study a possible matching silhouette first and snap a photo (to load into the Merlin Bird ID app later), then move closer to confirm your sighting. Once you've settled on a shrike, scan any surrounding barbed wire or long, pointy branches and plants for a chance to see one of the bird's well-butchered snacks. And stay a while (see tip #4) to watch it hunt for prey!

99

LOOK AROUND COWS AND HORSES

LET'S FACE IT, cows and horses are much easier to spot in the landscape than birds—whether by sight, sound . . . or smell. But they can also help you find more birds. Next time you pass by grazing cattle, or even just a horse or two, check the surrounding ground for a large white bird with a beautiful ginger-orange color on its head. Standing about 20 inches (51 cm) tall, the aptly named Cattle Egret eats the many flies that buzz around and alight on cows, and the many insects the cows kick up when grazing. (This seems like a much welcome service; a survey of Cattle Egrets in Okefenokee Swamp showed cattle exhibiting behaviors that encouraged the birds to forage at their side and, at times, on their backs.)

Cattle Egrets forage beside horse.

While Cattle Egrets are native to parts of Europe, Africa, and Asia, they have been expanding their range since the 1800s and have been reported on every continent. Unlike many other bird species, they have actually benefited from deforestation and the conversion of land to grazing pastures and croplands.

GOOD TO KNOW: *Cattle Egrets generally do not pick ticks off large mammals as do the oxpeckers of Africa, yet their Arabic name, Abu Qerdan, means "father of ticks."*

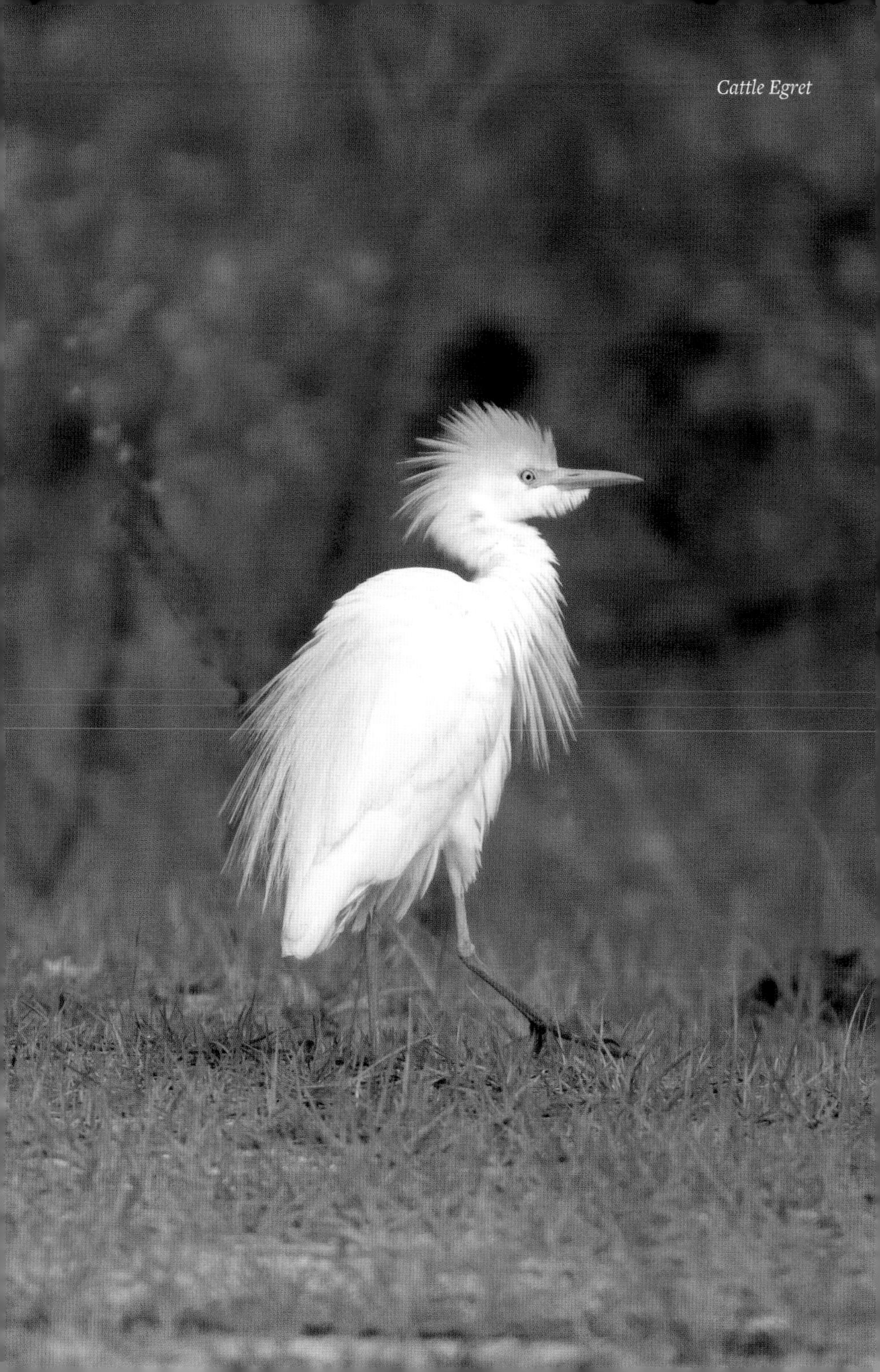
Cattle Egret

100

SANDERLINGS OF THE WORLD'S SHORELINES

IF YOU'VE EVER BEEN to any of the world's coastlines, you've likely seen them—the small black-billed shorebirds that run toward the water as the tide goes out, furiously digging their short bills into the sand, then running away from the approaching tide at the last possible second. These Sanderlings, a type of sandpiper, are probing for tiny mollusks and crustaceans that burrow into the sand.

It's hard to believe that every one of these birds hatched in the circumpolar Arctic; that's the only place in the world they breed. (There are even records of Sanderlings breeding on some of the landmasses closest to the North Pole, such as Cape Morris Jesup in Greenland.) Most Sanderlings in North America will return there each spring, though some (nonbreeding) will stick around and forgo the journey—it can be a long one, at least 1,800 miles (2,897 km). Sanderlings of South America have a tougher trek; many fly more than 6,000 miles (9,656 km) to lay their eggs on the tundra. It's no wonder that more of these birds, especially juveniles, choose to stay put in spring rather than migrate north. Either way, that's good news for finding Sanderlings; you can expect to see them scurrying along the beach year-round.

GOOD TO KNOW: *Shorebird species can be difficult to distinguish, but the Sanderling's bill provides helpful identification clues: It's black, straight, and noticeably shorter than those of other sandpipers. But it's the frenzied running in and out with the tide that makes it easy to confirm you've found yourself a Sanderling.*

Sanderling on Hilton Head Island, SC

ADVANCED BIRD-FINDING

101 KNOW YOUR TARGETS

FEEL LIKE SEEING A BIRD SPECIES you've never laid eyes on? eBird's "Targets" feature (eBird.org/targets) shows you nearby birds that you need for your life list. On a recent trip to Cupertino, California, I loaded up my targets and discovered I had a great chance of seeing a California Thrasher. Once I looked at photos of the bird—and its long, curved bill—I just *had* to meet it. A quick click showed me the many local hotspots where it was being seen; I chose one with consistent daily sightings of the bird: Ulistac Natural Area in Santa Clara. Taking a look at the location's recently submitted checklists and their attached photos, it was clear that the bird was usually spotted at dawn, singing from treetops. As I read more about the bird and its behavior, I learned that later in the day it often foraged low in dense chaparral shrubs and was difficult to find. To increase my chances of hitting my target, I would need to arrive by dawn and listen for it upon arrival. The plan was perfect; within a minute of stepping out of the car, I heard some incessant mimicry and followed it to find this California Thrasher. (Thrashers are in the Mimidae family, the same as mockingbirds.)

GOOD TO KNOW: *You can also view targets for birds you haven't seen this year, month, or day for the world or a selected region, as well as species for which you have yet to submit a photo or audio recording, aka "media lifers."*

California Thrasher

Northern Flicker

102
PRACTICE TRACKING BIRDS IN FLIGHT WITH COMMON BIRDS

GREAT FOR:

- Getting better views of birds when they land in a more open spot
- Locating nests
- Learning birds' different flight styles

HOW MANY TIMES HAVE you seen a cool-looking bird dash by, leaving you guessing as to what bird you *almost* found? Or maybe something passes by so fast you're not even sure it's a bird. Tracking flying objects is tough. If I were to stand at the receiving end of a professional tennis player's serve, I'd certainly miss. I might not even see the ball traveling toward me; it might just immediately arrive. (Professional ball players have high dynamic visual acuity, or DVA: the ability to detect details in a moving object, and they can better track objects in motion.)

Birds are fast, but not that fast. Tracking birds in flight is easy to master because you can practice any time of year on the most common birds in your area. The next time you see a bird fly by, try locking in on it with your naked eye and follow it as far as you can to see where it lands. You'll get better with practice, and it will eventually feel like the birds have slowed down. Also try it with your binoculars; it's tougher, but that's what your neighborhood birds can help with. Focus on larger birds or those in the distance at first; make it a game and you'll become a flight-tracking pro. Your improved DVA means you'll find more birds. Soon, you'll be able to easily focus on a tiny songbird darting by; you'll also notice that it has a caterpillar in its bill. Track it to find a nest full of baby birds.

GOOD TO KNOW: *A bird's flight pattern is a big clue to identifying it. Some common patterns are: direct, with straight flight and continuous wing-flapping (Great Blue Heron); flap-and-glide, similar to direct but with intermittent flapping (Tree Swallow); and undulating, with up-and-down movement through alternating flapping and gliding (sparrows and woodpeckers).*

103

EKE BIRDS OUT OF THE SKY/ PUT YOUR BINOCULARS TO THE TEST

GREAT FOR:

Distant raptors, vultures, flocks of birds

BELIEVE IT OR NOT, it is possible to eke birds out of the sky. You just need superhero eyes, or rather, a pair of bird-worthy binoculars (see tip #7). Some birds fly too high to see with the naked eye, especially during their migration. To us, it's as if they aren't even there to find. Bald Eagles, for example, can reach an altitude of 10,000 feet (3,048 m), and vultures often soar twice as high. Many other birds, including waterfowl and gulls, also reach amazing heights.

We often think of binoculars as a tool to get a better view of something we can already see, but when it comes to birding, they can also provide us with extra reach to find more birds in an otherwise empty sky. To find these high flyers, you'll need a point of reference on which to focus your binoculars: a very tall tree, building, or tower—even an airplane. As you focus on the object, slowly scan around and behind it for movement; your assisted eyes could land on a large raptor, or a tiny dot even farther away. If you see the latter, don't despair; soaring birds fly very fast, and whatever it is could come into better view in a matter of minutes. But enjoy the flight, because once you remove your binoculars from your eyes, the bird will vanish into thin air. Use the same technique to scan horizontally over treetops and rows of buildings, or vertically along towers or bridges. (Scan in a straight line to prevent dizziness!)

My fondest memory of eking a bird out of the sky was in the Walmart parking lot in St. Augustine. With binoculars to my eyes, I was admiring two Black Vultures soaring above when I noticed a small flying dot far behind them. I locked in on the moving speck until it flew in closer and I could tell what it was: a Mississippi Kite!

Black Vulture. Soaring vultures are great reference points to watch through your binoculars while looking for distant raptors. Planes work, too!

104 HEAD OUT AFTER A STORM

GREAT FOR:

- A chance to experience "bird fallout"
- Rare birds blown off course

YOU MAY HAVE HEARD the term "bird fallout," when thousands of migrating birds encounter extreme weather and winds and have to abandon their current travel plans; they basically fall out of the sky. This is a rare event; to experience it you either need to be in the right place at the right time or chase potential fallout from storms. I've only experienced a single fallout in more than ten years of birding, on the Gulf Coast of Florida after a major weather event. Dozens of wood warbler species, along with vireos, Yellow-billed Cuckoos, Scarlet Tanagers, and more, descended on my friend's backyard. Everywhere I looked, I saw birds; there were often too many to count. That's what you can expect in a fallout, and why so many birders do anything they can to get to one.

Even if you're not well positioned for a fallout, that doesn't mean you can't reap the benefits of storms. After Hurricane Isaac, over a hundred Magnificent Frigatebirds were temporarily stranded near where I lived. These birds spend most of their life in flight, yet there they were, forced to perch on trees along Pensacola Bay. Smaller storms and high winds often push pelagic birds—those that spend most of their life at sea—closer to shore. Walking along St. Augustine Beach in Florida on a moderately windy day, I noticed a small, dark bird flying in a gracefully erratic manner along the incoming waves. It was a storm-petrel, another pelagic bird usually found much farther offshore. Even uncommon nonpelagic birds might appear in inclement weather. Try to get out after a storm; you never know what birds you might see.

Magnificent Frigatebird

Ring-necked Pheasant

105

GET A "LITTLE" LOST

WHILE I NEVER WANT to repeat the time Connor and I got lost in Tate's Hell State Forest in Florida sans water or snacks (see tip #15), there's something to be said for getting a "little" lost. As we walked the back trails of Salt Marsh Nature Center in Brooklyn, we came upon a golf course where a Red-shouldered Hawk was hunting a squirrel that scurried to safety. We watched as the hawk struggled to relocate its prey, hungrily eyeing the golf greens from its perch a few feet away. But what we really wanted to see was a Ring-necked Pheasant: a large, ornate bird that was occasionally spotted in the area.

On our way back, the paths looked different than we remembered. Thankfully, being in the city, we were in no danger of being truly lost; we knew the general direction we needed to go and just couldn't find the right trail. After a few false starts on various side paths, Connor whispered, "There it is!" "What?" I asked. "The pheasant!" By the time I realized what he was talking about, the bird was out of sight. Using tip #4 ("Stay in one spot") and tip #13 ("Don't lose hope"), I meticulously scanned around the scrub for a dark green head and a red wattle. It took less than two minutes to find the beauty. For the very first time, I laid my eyes on its colorful headdress, intricate plumage, and ridiculously long tail. A life bird. And only because we'd gotten a little lost.

Juvenile Black-crowned Night-Heron

106

LOOK AT NIGHT

GREAT FOR:

- Night-herons
- Owls
- Migrating birds in well-lit parks during spring and fall

WAY BEFORE I KNEW birding was a thing, I found the most awesome bird at night in the quad of a university in North Carolina. My friend Harry and I heard a "Who cooks for you?" call and looked up to see an owl perched in the oak above us. At the time I had no idea it was a Barred Owl, but it was so exciting that we returned each night for weeks to find it—until it stopped hooting. (These owls vocalize mainly during breeding season.)

A well-lit campus, park, or sports field is a great place to look and listen for birds at night. You'll be surprised at how many types of birds are active after dark, including migratory songbirds. In fall and spring, birds are aiming to fatten up for their journey, so it makes sense that some will continue to snack at night. Look ahead on paths and their borders during your night walks for bobbing, strutting, or other birdy moves. Investigate bird sounds or songs. (Yes, some songbirds sing at night, including my favorite bird, the White-throated Sparrow.)

Great Egret

Your chances of finding birds at night increase near water sources, where you can detect movement and silhouettes of ducks, gulls, and herons against the water. One of my favorite things to do is find a foraging Black-crowned Night-Heron along Brooklyn Bridge Park's mini beach in late summer and fall. Great Blue Herons are also active at night and—luckily—have one of the easiest silhouettes to detect (a gigantic prehistoric-looking one). On your next night walk, be on the lookout and find more birds.

GOOD TO KNOW: *Safety first! Only go for solo night walks in familiar, safe areas, and always let someone know where you'll be. Even better, bring a friend along on your night walks.*

Ruby-crowned Kinglet

107

BIRD IN THE MIDDLE OF THE DAY

PEOPLE OFTEN ASK, "Do I have to get up early to find birds?" While that certainly is a common sentiment, likely propagated by phrases like "the early bird catches the worm" and guided bird walks that start at the crack of dawn, the answer is no. Some of your friends may be extreme birdwatchers who awake with the sun, but what you may not know is that many are looking for birds *all day long,* until the sun goes down! So you can just start looking later. Ducks, geese, loons, grebes, and shorebirds are out and about in their habitat throughout the day. Many hawks and vultures prefer afternoon outings when sun-warmed earth generates thermals—swirls of hot air they ride effortlessly.

Blue-headed Vireo

It's true that songbirds are often more active—and vocal—in the early morning, but during migration, you can find them at any time of day. When birds such as warblers, vireos, and thrushes prepare for their long journey, they go through hyperphagia: a state of extreme hunger and appetite that results in some of them doubling their body weight. These birds basically spend the whole day eating. One fall migration I had arrived at Brooklyn Bridge Park at 8:00 AM, but the birds didn't get hopping until much later, after 10:00 AM. I was suddenly surrounded by migrating birds devouring insect after insect. After spotting dozens of Ruby-crowned Kinglets and a Blue-headed Vireo foraging at high noon, I realized I needed to eat too and headed home for lunch.

GREAT FOR:

- Migrating songbirds fueling up for their journey
- Roosting night-herons in summer, perching motionless under the shade of the tree canopy

GOOD TO KNOW: *An hour before sunset is also a great time to find birds as they fill up their food stores until the next day breaks.*

108

STAY IN THE CAR

THE PESKY THING ABOUT BIRDS is that they fly away. So many times, you see one, take a tiny step toward it, and it's gone. But if you have a car, you have a very useful tool for bird observation: For some reason, birds aren't as bothered when we observe them from the confines of a car, which functions like the ultimate portable bird blind (a wall or structure you can hide behind to observe birds at close range).

If you come across a shape or silhouette of a bird while driving, and can safely stop or pull over, stay in the car at least for a few minutes to assess the situation. Opening the door right away and stepping outside could cause the bird—and other birds hidden but close by—to flee. Rolling down the windows is fine. Other birds along the side of the road and the surrounding area may start to approach. And if the bird that prompted you to stop was a raptor, you may even get to see it hunt.

While "car birding" with my friends Janet and Alan at a decommissioned airfield in Brooklyn, we came across a Snowy Owl bathing in a puddle on the road. As we slowly pulled to a stop and rolled down the windows for photos, the impressive bird stared at us and went back to its bath.

GOOD TO KNOW: *Bird-finding by car is best done on quiet back roads. Even from a car, maintain a respectful distance from birds; if a bird appears alarmed, view it from farther up the road. Some refuges, including Edwin B. Forsythe National Wildlife Refuge in New Jersey and Bear River Migratory Bird Refuge in Utah, have wildlife drives and loops that pass by fields, wetlands, and other birdy habitats.*

Snowy Owl

109

USE THE EYES IN THE BACK OF YOUR HEAD/ DETECT FLYING SHADOWS

GREAT FOR:

- Low-flying, large birds and flocks, especially vultures and Brown Pelicans
- Detecting bird movement in the trees above on sunny days

JUST LIKE ANY OTHER SKILL, finding birds gets easier the more you do it. You'll get used to their movements, mannerisms, shapes, and sounds—even their shadows. I discovered this a few years ago after the shadows cast by a flying flock of Cedar Waxwings caught my eye. I had found more birds by literally staring at the sidewalk.

To spot a flying shadow, both you and the bird need to be in the right position at the right time—and at the proper angle. With the sun to your back, its rays will be projected in front of you, where you'll see your own shadow. Any low-flying birds passing close to or through the line from you to the sun—behind or above you—will also cast a shadow in front of you. It's like having eyes in the back of your head!

The shadows that will lead you to the most birds are those cast on open areas like lawns, fields, and parking lots by large birds and flocks. You'll have plenty of time to notice them and react. When I was looking for birds in the wetlands of Harns Marsh in Fort Myers, Florida, every few minutes a large shadow was cast over the adjacent field by a low- and slow-flying Turkey or Black Vulture. This made me look up more often and find even more high flyers that were out of shadow range: Crested Caracara, Peregrine Falcon, Bald Eagle, and Sandhill Crane. From the patio of an Airbnb in St. Augustine, I often saw the shadows of Brown Pelican flocks on the parking lot, alerting me to their flyover a second later.

GOOD TO KNOW: *You can also use shadows to detect movement in a tree canopy. Look to the ground and check for movement among the shadows of the leaves. I often do this on sunny days to give my eyes a rest.*

Brown Pelicans cast shadows over a Florida parking lot.

A Lesser Black-backed Gull and Ring-billed Gulls

110 PLAY THE FLOCK GAME

IT'S EASY TO PASS by a foraging flock of House Sparrows daily without giving them a second look. Scanning dozens of LBJs—"little brown jobs"—on a lawn or tree might not seem like the most exciting bird-finding opportunity. But checking flocks is an easy way to find more birds, especially during spring and fall when many native sparrows

Look for something different; a rare Lesser Black-backed Gull roosts among Ring-billed Gulls.

and buntings migrate through our parks, yards, and playgrounds and forage in the same habitats as birds we see every day. So make a game of it! Systematically scan a flock with your binoculars, moving in vertical or horizontal rows and looking for something different: a longer tail, a different color or facial pattern, an odd behavior. This is a great way to find rare birds. I admit I was shocked to see a rare Painted Bunting perched among LBJs at the local House Sparrow hangout, atop a Brooklyn ice cream stand.

Play the flock game with shorebirds on the beach, gulls at their evening roost, or really any large group of birds in front of you. You might even whistle that old Muppets tune for inspiration: "One of these things is not like the other, one of these things does not belong. . . ."

GREAT FOR:

- Migrating sparrows and buntings
- A chance to find a Dickcissel, a beautifully patterned bunting often found foraging and flocking with House Sparrows during migration in the central and eastern US
- Finding rare gulls among a roost of common species

GOOD TO KNOW: *Distinguishing different sparrow species ("LBJs") may seem daunting at first, but it's easier than you think. Peruse the sparrow section of your field guide (see tip #6) to get familiar with the differences in head and facial patterns of the various species. Chest pattern is an important clue as well: streaked or plain, center dot or not, etc. During migration, one of these patterns will ring a bell as you're playing the flock game. Snap a photo for Merlin or compare it with the sparrows in your field guide.*

111 PREPARE FOR TAKEOFF TO FIND A RAPTOR

BIRDS FLEE EN MASSE for many reasons, such as an ear-splitting helicopter or a dog running toward them on a lawn, but one of the most common is the presence of a hawk, falcon, or owl. Anytime a group of birds suddenly takes off in flight, immediately scan above for a flying raptor. Also check surrounding posts, trees, and other perches.

Listen for "alarm calls" emanating from where the fearful flock settled. Many birds will activate such a red alert to warn others that the danger is present. Getting familiar with alarm calls is a great skill to add to your raptor-finding tool kit.

GOOD TO KNOW: *Different birds have different alarm calls. Many sound loud and agitated and are easy to detect. The Black-capped Chickadee, however, often uses its cute* chick-a-dee-dee-dee *call as an alarm but adds more* dees *on the end to indicate more danger. If you hear a chickadee doing way too many* dees, *check for a nearby predator.*

Red-shouldered Hawk

Resplendent Quetzal in Costa Rica

AS YOUR JOURNEY CONTINUES

I hope you've already started to discover birds in your neighborhood that you never knew were there. And once you find them, the obvious next step is to identify what you're seeing. If you're not sure of the species, it's tempting to jump straight into a field guide or app to start solving the mystery. But another option you might consider first is to spend quality time with the bird. Listen closely to its song, study its plumage patterns, and note what it's doing and the type of food it's eating. Is that what I did when I started birding? Certainly not! I'd immediately start looking through my field guide, trying to find whatever I'd just spotted. But as my head was buried in bird descriptions, I was missing out on the best part: being present with the bird in its habitat. By actively observing a bird, you'll connect with nature on a whole new level. And all of the things you notice about a bird will help you more easily find a match in your field guide once it leaves. Snapping a couple of reference photos is a great way to get back to observing a bird you can't identify and to save the sleuth-work for later.

As your bird story unfolds, I hope new dreams surface, too. One of mine was to see a Resplendent Quetzal, and it came true in the winter of 2023 in Costa Rica. I'm so excited that you, too, are living your birding adventure. A big part of the thrill is that we never know what might come next, whether it's finding a mega-rare bird in our backyard, seeing a crow clamming when all the trash cans are covered in snow, or taking a birding trip to a hotspot we just discovered. Here's to a lifetime of bird-finding!

BIBLIOGRAPHY

"ABA Code of Birding Ethics." American Birding Association, aba .org/aba-code-of-birding-ethics, accessed December 9, 2022.

Artuso, C., C. S. Houston, D. G. Smith, and C. Rohner, "Great Horned Owl (*Bubo virginianus*), version 1.0," in *Birds of the World*, ed. A. F. Poole (Ithaca, NY: Cornell Lab of Ornithology, 2020).

Eriksson, K., and H. M. Nummi, "Alcohol accumulation from ingested berries and alcohol metabolism in passerine birds," *Ornis Fennica* 60, no. 1 (1983): 2–9.

Foote, J. R., D. J. Mennill, L. M. Ratcliffe, and S. M. Smith, "Black-capped Chickadee (*Poecile atricapillus*), version 1.0," in *Birds of the World*, ed. A. F. Poole (Ithaca, NY: Cornell Lab of Ornithology, 2020).

Higashi, S., M. Ohara, H. Arai, and K. Matsuo, "Robber-like pollinators: overwintered queen bumblebees foraging on *Corydalis ambigua*," *Ecological Entomology* 13 (1988): 411–18.

Hilty, S., and D. A. Christie, "Bananaquit (*Coereba flaveola*), version 1.0," in *Birds of the World*, eds. J. del Hoyo, A. Elliott, J. Sargatal, D. A. Christie, and E. de Juana (Ithaca, NY: Cornell Lab of Ornithology, 2020).

Levey, D. J., and G. E. Duke, "How Do Frugivores Process Fruit? Gastrointestinal Transit and Glucose Absorption in Cedar Waxwings (*Bombycilla cedrorum*)," *The Auk* 109, no. 4 (October 1992): 722–30.

Macwhirter, R. B., P. Austin-Smith Jr., and D. E. Kroodsma, "Sanderling (*Calidris alba*), version 1.0," in *Birds of the World*, ed. S. M. Billerman (Ithaca, NY: Cornell Lab of Ornithology, 2020).

Smallwood, J. A., and D. M. Bird, "American Kestrel (*Falco sparverius*), version 1.0," in *Birds of the World*, eds. A. F. Poole and F. B. Gill (Ithaca, NY: Cornell Lab of Ornithology, 2020).

Otis, D. L., J. H. Schulz, D. Miller, R. E. Mirarchi, and T. S. Baskett, "Mourning Dove (*Zenaida macroura*), version 1.0," in *Birds of the World*, ed. A. F. Poole (Ithaca, NY: Cornell Lab of Ornithology, 2020).

Tallamy, D. W., *Nature's Best Hope: A New Approach to Conservation That Starts in Your Yard* (Portland, OR: Timber Press, 2020).

Uchida, Y., D. Kudoh, A. Murakami, M. Honda, S. Kitazawa, "Origins of Superior Dynamic Visual Acuity in Baseball Players: Superior Eye Movements or Superior Image Processing," *PLOS ONE* 7, no. 2 (2012): e31530.

ACKNOWLEDGMENTS

To the birds that continue to amaze me with their feats. I could not have written this book without them.

To my mentors, Bob and Lucy Duncan, for getting me hooked on birds in Florida and continuing to support and inspire me.

To my colleagues past and present at the Cornell Lab of Ornithology, especially those on the eBird and Merlin teams whose incredible bird-finding tools are mentioned throughout the book.

To Greg Wolf, Laura Dobell, William Grayson Wolf, and Isabelle Wolf for their enthusiastic support and especially for letting me use their beautiful urban terrace to write.

To Bets Radley, Eric Engleman, and Kim Argetsinger for reviewing the list of tips and providing such helpful feedback and ideas.

To Harold Moeller for generous feedback on photos and text.

To Kathryn Heintz for looking over the book and offering such helpful advice and suggestions.

To Janet and Richard Eargle for use of an inspiring outdoor writing space with so many different birds—and American beavers—to see. My favorite accounts were written in their yard.

To John Connor, for coming up with the ideas for the sections, reviewing the manuscript, and putting the "detect flying shadows" tip to the test.

To Allison Dubinsky for an incredible proofread that exceeded all of my expectations.

To the inspiring and talented team at The Experiment, especially my editor, Nick Cizek, designer Jack Dunnington, and Matthew Lore. Many thanks for planting the seeds of the idea for such a fun book!

ABOUT THE AUTHOR

HEATHER WOLF is a Brooklyn-based birder, author, photographer, and educator. She works with the Cornell Lab of Ornithology as a web developer, teaches birding classes at Brooklyn Botanic Garden, and gives walks and talks for various organizations in New York City and beyond. She is also the author of *Birding at the Bridge: In Search of Every Bird on the Brooklyn Waterfront.*

heatherwolf.com | **heatherwolf** | **realheatherwolf**